Industrial Ecology

Industrial Ecology

Edited by
Liam Spencer

www.larsen-keller.com

Industrial Ecology
Edited by Liam Spencer
ISBN: 978-1-63549-695-6 (Hardback)

© 2018 Larsen & Keller

Published by Larsen and Keller Education,
5 Penn Plaza,
19th Floor,
New York, NY 10001, USA

Cataloging-in-Publication Data

Industrial ecology / edited by Liam Spencer.
    p. cm.
Includes bibliographical references and index.
ISBN 978-1-63549-695-6
1. Industrial ecology. 2. Materials management. 3. Environmental management.
I. Spencer, Liam.
TS161 .I53 2018
658.7--dc23

This book contains information obtained from authentic and highly regarded sources. All chapters are published with permission under the Creative Commons Attribution Share Alike License or equivalent. A wide variety of references are listed. Permissions and sources are indicated; for detailed attributions, please refer to the permissions page. Reasonable efforts have been made to publish reliable data and information, but the authors, editors and publisher cannot assume any responsibility for the vailidity of all materials or the consequences of their use.

Trademark Notice: All trademarks used herein are the property of their respective owners. The use of any trademark in this text does not vest in the author or publisher any trademark ownership rights in such trademarks, nor does the use of such trademarks imply any affiliation with or endorsement of this book by such owners.

For more information regarding Larsen and Keller Education and its products, please visit the publisher's website www.larsen-keller.com

Table of Contents

Preface   VII

Chapter 1   **An Introduction to Industrial Ecology**   1
- i. Industrial Ecology   1
- ii. Industrial Symbiosis   12

Chapter 2   **Various Concepts Related to Industrial Ecology**   16
- i. Life-cycle Assessment   16
- ii. Design for the Environment   28
- iii. Extended Producer Responsibility   33
- iv. Helix of Sustainability   40

Chapter 3   **Industrial Ecology: Innovative Approaches**   44
- i. Ecomechatronics   44
- ii. Cradle-to-cradle Design   47
- iii. Ecodesign   56
- iv. Integrated Chain Management   62
- v. Environmental Management System   63
- vi. Ecological Modernization   66
- vii. Regenerative Design   69

Chapter 4   **Methods and Techniques of Industrial Ecology**   73
- i. SWOT Analysis   73
- ii. Zero Waste   81
- iii. Eco-costs Value Ratio   93
- iv. Ecological Footprint   100
- v. Energy Accounting   107
- vi. Rebound Effect (Conservation)   108
- vii. Environmental Full-cost Accounting   114

Chapter 5   **Integrating Industrial Ecology and Economy**   120
- i. Dematerialization (Economics)   120
- ii. Industrial Organization   120
- iii. Circular Economy   123
- iv. Eco-efficiency   132
- v. Ecoleasing   135
- vi. Eco-investing   136

Chapter 6 **Materials Flow in Industrial Ecology** — **139**
- i. Material Flow — 139
- ii. Material Flow Analysis — 140
- iii. Plant Simulation — 147
- iv. AnyLogic — 151
- v. Supply Chain Management — 160
- vi. Industrial Metabolism — 176
- vii. Reuse — 177
- viii. Remanufacturing — 187
- ix. Recycling — 192

Permissions

Index

Preface

Industrial ecology refers to the research and practice of energy and material flows in an industrial system. It also includes processes which transform resources present in the nature into something useful for the human needs. The subject combines elements of engineering, toxicology, natural sciences, economics, and sociology to understand the importance of natural elements and their usefulness to human beings. While understanding the long-term perspectives of the topics, the textbook makes an effort in highlighting their impact as a modern tool for the growth of the discipline. It will serve as a valuable source of reference for those interested in this area. This textbook aims to assist those with a goal of delving into the vast field of industrial ecology.

Given below is the chapter wise description of the book:

Chapter 1- Industrial ecology focuses on shifting industrial processes from open loop systems to closed loop system so that no material remains as a waste product. The subject focuses on certain areas are material and energy flow studies, life-cycle planning, eco-efficiency, design and assessment, etc. The chapter on industrial ecology offers an insightful focus, keeping in mind the complex subject matter.

Chapter 2- The various concepts related to industrial ecology are life-cycle assessment, design for the environment, extended producer responsibility and helix of sustainability. The technique that is used to evaluate environmental impacts that are related with the stages of a product's life is known as life-cycle assessment. This chapter is an overview of the subject matter incorporating all the major aspects of industrial ecology.

Chapter 3- The application of mechatronical technology to reduce the cost of ownership of machines and the ecological impact they have is known as ecomechatronics. Cradle-to-cradle design, integrated chain management, environmental management system, ecodesign, ecological modernization and regenerative design are some significant and important topics related to industrial ecology. The following chapter unfolds its crucial aspects in a critical yet systematic manner.

Chapter 4- The methods and techniques of industrial ecology are SWOT analysis, ecological footprint, energy accounting, zero waste, eco-costs value ratio and rebound effect. SWOT analysis is carried out for companies and products. It can be categorized into internal factors and external factors. This chapter discusses the methods of industrial ecology in a critical manner providing key analysis to the subject matter.

Chapter 5- Dematerialization is the cutting down of the quantity of materials that would be used by society. Industrial organization, circular economy, eco-efficiency, etc. are some important topics related to the subject of industrial ecology and economy. This chapter helps the readers in developing a better idea about industrial ecology and economy.

Chapter 6- The transportation of raw materials and other vital industrial elements from one place to another is known as material flow. The typical tools used in the process are AnyLogic, plant simulation for production systems and AutoMod for logistics systems. Other themes include supply chain management, industrial metabolism, etc. This chapter will provide an integrated understanding of materials flow in industrial ecology.

At the end, I would like to thank all those who dedicated their time and efforts for the successful completion of this book. I also wish to convey my gratitude towards my friends and family who supported me at every step.

Editor

An Introduction to Industrial Ecology

Industrial ecology focuses on shifting industrial processes from open loop systems to closed loop system so that no material remains as a waste product. The subject focuses on certain areas are material and energy flow studies, life-cycle planning, eco-efficiency, design and assessment, etc. The chapter on industrial ecology offers an insightful focus, keeping in mind the complex subject matter.

Industrial Ecology

Industrial ecology (IE) is the study of material and energy flows through industrial systems. The global industrial economy can be modelled as a network of industrial processes that extract resources from the Earth and transform those resources into commodities which can be bought and sold to meet the needs of humanity. Industrial ecology seeks to quantify the material flows and document the industrial processes that make modern society function. Industrial ecologists are often concerned with the impacts that industrial activities have on the environment, with use of the planet's supply of natural resources, and with problems of waste disposal. Industrial ecology is a young but growing multidisciplinary field of research which combines aspects of engineering, economics, sociology, toxicology and the natural sciences.

Industrial ecology has been defined as a "systems-based, multidisciplinary discourse that seeks to understand emergent behaviour of complex integrated human/natural systems". The field approaches issues of sustainability by examining problems from multiple perspectives, usually involving aspects of sociology, the environment, economy and technology. The name comes from the idea that the analogy of natural systems should be used as an aid in understanding how to design sustainable industrial systems.

Overview

Industrial ecology is concerned with the shifting of industrial process from linear (open loop) systems, in which resource and capital investments move through the system to become waste, to a closed loop system where wastes can become inputs for new processes.

Much of the research focuses on the following areas:

- material and energy flow studies ("industrial metabolism")

- dematerialization and decarbonization
- technological change and the environment
- life-cycle planning, design and assessment
- design for the environment ("eco-design")
- extended producer responsibility ("product stewardship")
- eco-industrial parks ("industrial symbiosis")
- product-oriented environmental policy
- eco-efficiency

Example of Industrial Symbiosis. Waste steam from a waste incinerator (right) is piped to an ethanol plant (left) where it is used as in input to their production process.

Industrial ecology seeks to understand the way in which industrial systems (for example a factory, an ecoregion, or national or global economy) interact with the biosphere. Natural ecosystems provide a metaphor for understanding how different parts of industrial systems interact with one another, in an "ecosystem" based on resources and infrastructural capital rather than on natural capital. It seeks to exploit the idea that natural systems do not have waste in them to inspire sustainable design.

Along with more general energy conservation and material conservation goals, and redefining commodity markets and product stewardship relations strictly as a service economy, industrial ecology is one of the four objectives of Natural Capitalism. This strategy discourages forms of amoral purchasing arising from ignorance of what goes on at a distance and implies a political economy that values natural capital highly and relies on more instructional capital to design and maintain each unique industrial ecology.

History

View of Kalundborg Eco-industrial Park

Industrial ecology was popularized in 1989 in a *Scientific American* article by Robert Frosch and Nicholas E. Gallopoulos. Frosch and Gallopoulos' vision was "why would not our industrial system behave like an ecosystem, where the wastes of a species may be resource to another species? Why would not the outputs of an industry be the inputs of another, thus reducing use of raw materials, pollution, and saving on waste treatment?" A notable example resides in a Danish industrial park in the city of Kalundborg. Here several linkages of byproducts and waste heat can be found between numerous entities such as a large power plant, an oil refinery, a pharmaceutical plant, a plasterboard factory, an enzyme manufacturer, a waste company and the city itself. Another example is the Rantasalmi EIP in Rantasalmi, Finland. While this country has had previous organically formed EIP's, the park at Rantasalmi is Finland's first planned EIP.

The scientific field Industrial Ecology has grown quickly in recent years. The Journal of Industrial Ecology (since 1997), the International Society for Industrial Ecology (since 2001), and the journal Progress in Industrial Ecology (since 2004) give Industrial Ecology a strong and dynamic position in the international scientific community. Industrial Ecology principles are also emerging in various policy realms such as the concept of the Circular Economy that is being promoted in China. Although the definition of the Circular Economy has yet to be formalized, generally the focus is on strategies such as creating a circular flow of materials, and cascading energy flows. An example of this would be using waste heat from one process to run another process that requires a lower temperature. The hope is that strategy such as this will create a more efficient economy with fewer pollutants and other unwanted by-products.

Principles

One of the central principles of Industrial Ecology is the view that societal and technological systems are bounded within the biosphere, and do not exist outside of it. Ecology is used as a *metaphor* due to the observation that natural systems reuse materials and have a largely closed loop cycling of nutrients. Industrial Ecology approaches problems with the hypothesis that by using similar principles as *natural systems, industrial systems* can be improved to reduce their impact on the natural environment as well. The table shows the general metaphor.

Biosphere	Technosphere
• Environment	• Market
• Organism	• Company
• Natural Product	• Industrial Product
• Natural Selection	• Competition
• Ecosystem	• Eco-Industrial Park
• Ecological Niche	• Market Niche
• Anabolism / Catabolism	• Manufacturing / Waste Management
• Mutation and Selection	• Design for Environment
• Succession	• Economic Growth
• Adaptation	• Innovation
• Food Web	• Product Life Cycle

IE examines societal issues and their relationship with both technical systems and the environment. Through this *holistic view*, IE recognizes that solving problems must involve understanding the connections that exist between these systems, various aspects cannot be viewed in isolation. Often changes in one part of the overall system can propagate and cause changes in another part. Thus, you can only understand a problem if you look at its parts in relation to the whole. Based on this framework, IE looks at environmental issues with a *systems thinking* approach. A good IE example with these societal impacts can be found at the Blue Lagoon in Iceland. The Lagoon uses super-heated water from a local geothermal power plant to fill mineral-rich basins that have become recreational healing centers. In this sense the industrial process of energy production uses its wastewater to provide a crucial resource for the dependent recreational industry.

Take a city for instance. A city can be divided into commercial areas, residential areas, offices, services, infrastructures, and so forth. These are all sub-systems of the 'big city' system. Problems can emerge in one sub-system, but the solution has to be global. Let's say the price of housing is rising dramatically because there is too high a demand for housing. One solution would be to build new houses, but this will lead to more people living in the city, leading to the need for more infrastructure like roads, schools, more supermarkets, etc. This system is a simplified interpretation of reality whose behaviors can be 'predicted'.

In many cases, the systems IE deals with are complex systems. Complexity makes it difficult to understand the behavior of the system and may lead to rebound effects. Due to unforeseen behavioral change of users or consumers, a measure taken to improve environmental performance does not lead to any improvement or may even worsen the situation.

Moreover, *life cycle thinking* is also a very important principle in industrial ecology. It implies that all environmental impacts caused by a product, system, or project during its life cycle are taken into account. In this context life cycle includes

- Raw material extraction
- Material processing
- Manufacture
- Use
- Maintenance
- Disposal

The transport necessary between these stages is also taken into account as well as, if relevant, extra stages such as reuse, remanufacture, and recycle. Adopting a life cycle approach is essential to avoid shifting environmental impacts from one life cycle stage to another. This is commonly referred to as problem shifting. For instance, during the re-design of a product, one can choose to reduce its weight, thereby decreasing use of resources. However, it is possible that the lighter materials used in the new product will be more difficult to dispose of. The environmental impacts of the product gained during the extraction phase are shifted to the disposal phase. Overall environmental improvements are thus null.

A final important principle of IE is its *integrated approach* or *multidisciplinarity*. IE takes into account three different disciplines: social sciences (including economics), technical sciences and environmental sciences. The challenge is to merge them into a single approach.

Examples

The Kalundborg industrial park is located in Denmark. This industrial park is special because companies reuse each other's waste (which then becomes by-products). For example, the Energy E2 Asnæs Power Station produces gypsum as a by-product of the electricity generation process; this gypsum becomes a resource for the BPB Gyproc A/S which produces plasterboards. This is one example of a system inspired by the biosphere-technosphere metaphor: in ecosystems, the waste from one organism is used as inputs to other organisms; in industrial systems, waste from a company is used as a resource by others.

Apart from the direct benefit of incorporating waste into the loop, the use of an eco-industrial park can be a means of making renewable energy generating plants, like Solar PV, more economical and environmentally friendly. In essence, this assists the growth of the renewable energy industry and the environmental benefits that come with replacing fossil-fuels.

Additional examples of industrial ecology include:

- Substituting the fly ash byproduct of coal burning practices for cement in concrete production

- Using second generation biofuels. An example of this is converting grease or cooking oil to biodiesels to fuel vehicles.

- South Africa's National Cleaner Production Center (NCPC) was created in order to make the region's industries more efficient in terms of materials. Results of the use of sustainable methods will include lowered energy costs and improved waste management. The program assesses existing companies to implement change.

Tools

People	Planet	Profit	Modeling
• Stakeholder analysis • Strength Weakness Opportunities Threats Analysis (SWOT Analysis) • Ecolabelling • ISO 14000 • Environmental management system (EMS) • Integrated chain management (ICM) • Technology assessment	• Environmental impact assessment (EIA) • Input-output analysis (IOA) • Life-cycle assessment (LCA) • Material flow analysis (MFA) • Substance flow analysis (SFA) • MET Matrix	• Cost benefit analysis (CBA) • Full cost accounting (FCA) • Life cycle costing (LCC)	• Stock and flow analysis • Agent based modeling

Future Directions

The ecosystem metaphor popularized by Frosch and Gallopoulos has been a valuable creative tool for helping researchers look for novel solutions to difficult problems. Recently, it has been pointed out that this metaphor is based largely on a model of classical ecology, and that advancements in understanding ecology based on complexity science have been made by researchers such as C. S. Holling, James J. Kay, and further advanced in terms of contemporary ecology by others. For industrial ecology, this may mean a shift from a more mechanistic view of systems, to one where sustainability is viewed as an emergent property of a complex system. To explore this further, several researchers are working with agent based modeling techniques .

Exergy analysis is performed in the field of industrial ecology to use energy more efficiently. The term *exergy* was coined by Zoran Rant in 1956, but the concept was developed by J. Willard Gibbs. In recent decades, utilization of exergy has spread outside of physics and engineering to the fields of industrial ecology, ecological economics, systems ecology, and energetics.

Other Examples

Another great example of industrial ecology both in practice and in potential is the Burnside Cleaner Production Centre in Burnside, Nova Scotia. They play a role in facilitating the 'greening' of over 1200 businesses that are located in Burnside, Eastern Canada's largest industrial park. The creation of waste exchange is a big part of what they work towards, which will promote strong industrial ecology relationships.

History of Industrial Ecology

The establishment of industrial ecology as field of scientific research is commonly attributed to an article devoted to industrial ecosystems, written by Frosch and Gallopoulos, which appeared in a 1989 special issue of Scientific American. Industrial ecology emerged from several earlier ideas and concepts, some of which date back to the 19th century.

Before the 1960s

The term "industrial ecology" has been used alongside "industrial symbiosis" at least since the 1940s. Economic geography was perhaps one of the first fields to use these terms. For example, in an article published in 1947, George T. Renner refers to "The General Principle of Industrial Location" as a "Law of Industrial Ecology". Briefly stated this is:

Any industry tends to locate at a point which provides optimum access to its ingredients or component elements. If all these component elements be juxtaposed, the location of the industry is predetermined. If, however, they occur widely separated, the industry is so located as to be most accessible to that element which would be the most expensive or difficult to transport and which, therefore, becomes the locative factor for the industry in question.

In the same article the author defines and describes industrial symbiosis:

Often the location of an industry cannot be fully understood solely in terms of its locative ingredient elements. There are relationships between industries, sometimes simple, but often quite complex, which enter into and complicate the analysis. Chief among these is the phenomenon of industrial symbiosis. By this is meant the consorting together of two or more of dissimilar industries. Industrial Symbiosis, when scrutinized, is seen to be of two kinds, disjunctive and conjunctive.

It appears that the concept of Industrial Symbiosis was not new for the field of economic geography, since the same categorization is used by Walter G. Lezius in his 1937 article "Geography of Glass Manufacture at Toledo, Ohio", also published in the Journal of Economic Geography.

Used in a different context, the term "Industrial Ecology" is also found in a 1958 paper concerned with the relationship between the ecological impact from increasing urbanization and value orientations of related peoples. The case study is in Lebanon:

The central ecological variable in the present research is ecological mobility, or the movement of men in space. It is patent that modern Industrial Ecology requires more such adaptive mobility than does traditional folk-village organization.

1960s

In 1963, we find the term Industrial Ecology (defined as the "complex ecology of the modern industrial world") being used to describe the social nature and complexity of (and within) industrial systems:

industrial organisations are social rather than mechanical systems. A firm is not only a working organisation with a working purpose. It is rather a community with its own 'politics', in so far as it is involved in problems concerned with the proper distribution of power between individuals and groups of individuals and with questions of individual and group prestige, influence, status and standing [and he concludes that] the understanding which the student of management is expected to gain is no less than the attainment of insight into an Industrial Ecology of great complexity.

In 1967, the President of the American association for the advancement of science writes in "The experimental city" that "There are examples of industrial symbiosis where one industry feeds off, or at least neutralizes, the wastes of another" The same author in 1970 talks about "The Next Industrial Revolution" The concept of material and energy sharing and reuse is central to his proposal for a new industrial revolution and he cites agro-industrial symbiosis as a practical way for achieving this:

The object of the next industrial revolution is to ensure that there will be no such thing as waste, on the basis that waste is simply some substance that we do not yet have the wit to use The next industrial revolution is this generating of a huge new [industry that will not produce products, it will rather reprocess the things we call wastes so they may be reproduced in the factories into the things we need Having the city near the rural area will enable waste heat to be used to speed up the biological processes of treating the organic wastes before they go back into the land. This might end in an elegant arrangement-the power plants located close enough to the center of use, to the people who need the power, but also, within the economics, close enough to the agriculture lands so that the waste heat may be used there. This is an example of agro-industrial symbiosis, if you like to call it that.

In these early articles, "Industrial Ecology" is used in its literal sense - as a system of interacting industrial entities. The relation to natural ecosystems (through either metaphor or analogy) is not explicit. Industrial Symbiosis on the other hand, is already clearly defined as a type of industrial organization, and the term symbiosis is borrowed from the ecological sciences to describe an analogous phenomenon in industrial systems.

1970s

Industrial Ecology has been a research subject of the Japan Industrial Policy Research Institute since 1971. Their definition of Industrial Ecology is "research for the prospect of dynamic harmonization between human activities and nature by a systems approach based upon ecology (JIPRI, 1983)". This programme has resulted to a number of reports that are available only in Japanese.

One of the earliest definitions of Industrial Ecology was proposed by Harry Zvi Evan at a seminar of the Economic Commission of Europe in Warsaw (Poland) in 1973 (an article was subsequently published by Evan in the Journal for International Labour Review in 1974 vol. 110 (3), pp. 219–233). Evan defined Industrial Ecology as a systematic analysis of industrial operations including factors like: Technology, environment, natural resources, bio-medical aspects, institutional and legal matters as well as the socio-economic aspects.

In 1974 the term of Industrial Ecology is perhaps for the first time associated with a cyclical production mode (rather than a linear one, resulting to waste). In this article, the necessity for a transition to an "open-world Industrial Ecology", is used as argument for the need to establish lunar industries:

Low living standards provide one strong motive for most developing countries to increase their productivity and grow economically. Population increase is a still more powerful driver for increased world consumption. Thus the pressure on resources will continue to grow. Instead of deploring it, we better grow with it. Only through transition to an open-world Industrial Ecology - which includes both benign industrial revolution on Earth and extraterrestrial industrialization - can the present apparent limits to growth be overcome.

Many elements of modern Industrial Ecology were commonplace in the industrial sectors of the former Soviet Union. For example, "kombinirovanaia produksia" (combined production) was present from the earliest years of the Soviet Union and was instrumental in shaping the patterns of Soviet industrialization. "Bezotkhodnoyi tekhnologii" (waste-free technology) was introduced in the final decades of the USSR as a way to increase industrial production while limiting environmental impact. Fiodor Davitaya, a Soviet scientist from the Republic of Georgia, described in 1977 the analogy relating industrial systems to natural systems as a model for a desirable transition to cleaner production:

Nature operates without any waste products. What is rejected by some organisms provides food for others. The organisation of industry on this principle—with the waste products of some branches of industry providing raw material for others—means in effect using natural processes as a model, for in them the resolution of all arising contradictions is the motive force of progress.

1980s

By the 80s Industrial Ecology was already "promoted" to a research subject, which several institutes around the globe embraced. In a 1986 article published in the Journal of Ecological Modeling, there is a full description of Industrial Ecology and the analogy to natural ecosystems is clearly stated:

The structure and inner-working of an industrial society resemble those of a natural ecosystem. The concepts in ecology such as habitat, succession, trophic level, limiting factors and community metabolism can also apply to the study of the ecology of an industrial society. For instance, an industry in a society may grow or decline as a consequence of dynamic changes in exogenous limiting resources and in the hierarchical and/or metabolic structure of that society. When studying the ecology of an industrial society (henceforth termed 'Industrial Ecology'), these concepts and methodologies employed in ecosystems analyses are useful.

In fact, in the above article there is an attempt to model an "industrial ecological system". The model is composed of seven major sections: industry, population, labor force, living state, environment and pollution, general health, and occupational health. Notice the rough similarity with Evan's factors as stated in the above section.

During the 80s the emergence of another related term, "industrial metabolism", is observed. The term is used as a metaphor for the organization and functioning of industrial activity. In an article defending the "biological modulation of terrestrial carbon cycle", the author includes an extraordinary parenthetical note:

Parenthetically, it should be noted that it is an intrinsic property of life to proliferate exponentially until the encounter of limits set by (1) the availability of biologically utilizable reducing power, or (2) the exhaustion of some critical nutrient, or (3) an autotoxic effect imposed by life on its own environment. These limits are universal, applying to microbial ecosystems as well as to the population dynamics of a seemingly unrestricted biological superdominant such as Homo Sapiens (here, the ultimate limit is likely to be placed by an autotoxic effect exerted by the "extrasomatic" (industrial) metabolism of the human race).

1989 – Decisive Articles

In 1989 two articles were released that played a decisive role in the history of industrial ecology. The first one was titled "Industrial Metabolism" by Robert Ayres. Ayres essentially lays the foundations of Industrial Ecology, although the term is not to be found in this article. In the appendix of the article he includes "a theoretical exploration of the biosphere and the industrial economy as material-transformation systems and lessons that might be learned from their comparison". He proposes that:

We may think of both the biosphere and the industrial economy as systems for the transformation of materials. The biosphere as it now exists is nearly a perfect system for recycling materials. This was not the case when life on earth began. The industrial system of today resembles the earliest stage of biological evolution, when the most primitive living organisms obtained their energy from a stock of organic molecules accumulated during prebiotic times. It is increasingly urgent for us to learn from the biosphere and modify our industrial metabolism, the energy - and value - yielding process essential to economic development we should not only postulate, but indeed endorse, a long-run imperative favoring an industrial metabolism that results in reduced extraction of virgin materials, reduced loss of waste materials, and increased recycling of useful ones.

The term "Industrial Ecology" gains mainstream attention later the same year (1989) through a "Scientific American" article named "Strategies for Manufacturing". In this article, R.Frosch and N.Gallopoulos wonder "why would not our industrial system behave like an ecosystem, where the wastes of a species may be resource to another species? Why would not the outputs of an industry be the inputs of another, thus reducing use of raw materials, pollution, and saving on waste treatment?"

This vision gave birth to the concept of the Eco-industrial Park, the industrial complex that is governed by Industrial Ecology principles. A notable example resides in a Danish industrial park in the city of Kalundborg. There, several linkages of byproducts and waste heat can be found between numerous entities such as a large power plant, an oil refinery, a pharmaceutical plant, a plasterboard factory, an enzyme manufacturer, a waste company and the city itself.

Frosch's and Gallopoulos' thinking was in certain ways simply an extension of earlier ideas, such as the efficiency and waste-reduction thinking annunciated by Buckminster Fuller and his students (e.g., J. Baldwin), and parallel ideas about energy cogeneration, such as those of Amory Lovins and the Rocky Mountain Institute.

1990s

In 1991, C. Kumar Patel organized a seminal colloquium on Industrial Ecology, held on May 20 and 21, 1991, at the National Academy of Sciences in Washington D.C. The papers were later published in the Proceedings of the National Academy of Sciences USA, and they form an excellent reference on Industrial Ecology. Papers include

- "Industrial Ecology: Concepts and Approaches"
- "Industrial Ecology: A Philosophical Introduction"
- "The Ecology of Markets,"
- "Industrial Ecology: Reflections on a Colloquium"

All twenty three papers are available online.

21st Century

The *Journal of Industrial Ecology* (since 1997), the International Society for Industrial Ecology (since 2001), and the journal *Progress in Industrial Ecology* (since 2004) have covered industrial ecology in the international scientific community. Principles of industrial ecology are also emerging in various policy realms such as the concept of the circular economy that is being promoted in China. Although the definition of the circular economy has yet to be formalized, generally the focus is on strategies such as creating a circular flow of materials, and cascading energy flows. An example of this would be using waste heat from one process to run another process that requires a lower temperature. This maximizes the efficiency of exergy use. This strategy aims for a more efficient economy with fewer pollutants and other unwanted by-products.

Industrial Symbiosis

Industrial Symbiosis a subset of industrial ecology. It describes how a network of diverse organizations can foster eco-innovation and long-term culture change, create and share mutually profitable transactions - and improve business and technical processes.

Although geographic proximity is often associated with industrial symbiosis, it is neither necessary nor sufficient—nor is a singular focus on physical resource exchange. In practice, using industrial symbiosis as an approach to commercial operations – using, recovering and redirecting resources for reuse – results in resources remaining in productive use in the economy for longer. This in turn creates business opportunities, reduces demands on the earth's resources, and provides a stepping-stone towards creating a circular economy. The industrial symbiosis model devised and managed by International Synergies Limited is a facilitated model operating at the national scale in the United Kingdom (NISP - National Industrial Symbiosis Programme), and at other scales around the world. International Synergies Limited has developed global expertise in IS, instigating programmes in Belgium, Brazil, Canada, China, Denmark, Finland, Hungary, Italy, Mexico, Poland, Romania, Slovakia, South Africa and Turkey, as well as the UK. Industrial symbiosis is a subset of industrial ecology, with a particular focus on material and energy exchange. Industrial ecology is a relatively new field that is based on a natural paradigm, claiming that an industrial ecosystem may behave in a similar way to the natural ecosystem wherein everything gets recycled.

Introduction

Eco-industrial development is one of the ways in which industrial ecology contributes to the integration of economic growth and environmental protection. Some of the examples of eco-industrial development are:

- Circular economy (single material and/or energy exchange)
- Greenfield eco-industrial development (geographically confined space)
- Brownfield eco-industrial development (geographically confined space)
- Eco-industrial network (no strict requirement of geographical proximity)
- Virtual eco-industrial network (networks spread in large areas e.g. regional network)
- Networked Eco-industrial System (macro level developments with links across regions)

"This classification omits any industrial sector-wide approaches and appreciates the diversity of the industrial system which is a key feature of industrial symbiosis. It is aimed to include initiatives that focus on achieving utility sharing and symbiosis among diverse sectors of industry". It is the diversity and the openness of industrial symbiosis that makes it a unique approach to eco-industrial development.

Industrial symbiosis engages traditionally separate industries in a collective approach to competitive advantage involving physical exchange of materials, energy, water, and/or by-products. The keys to industrial symbiosis are collaboration and the synergistic possibilities offered by geographic proximity". The sharing of information is even more critical with the emergence of virtual globes such as Google Earth. These tools can greatly simplify the geographical analysis involved in determining potential IS opportunities.

Industrial symbiosis systems collectively optimize material and energy use at efficiencies beyond those achievable by any individual process alone. IS systems such as the web of materials and energy exchanges among companies in Kalundborg, Denmark have spontaneously evolved from a series of micro innovations over a long time scale; however, the engineered design and implementation of such systems from a macro planner's perspective, on a relatively short time scale, proves challenging. Nevertheless, there are examples of industrial symbiosis being approached as national / regional initiatives with some significant success particularly in Europe.

Often, access to information on available by-products is non-existent. These by-products are considered waste and typically not traded or listed on any type of exchange.

Example

- Recent work reviewed government policies necessary to construct a multi-gigaWatt photovoltaic factory and complementary policies to protect existing solar companies are outlined and the technical requirements for a symbiotic industrial system are explored to increase the manufacturing efficiency while improving the environmental impact of solar photovoltaic cells. The results of the analysis show that an eight-factory industrial symbiotic system can be viewed as a medium-term investment by any government, which will not only obtain direct financial return, but also an improved global environment. This is because synergies have been identified for co-locating glass manufacturing and photovoltaic manufacturing. The waste heat from glass manufacturing can be used in industrial-sized greenhouses for food production. Even within the PV plant itself a secondary chemical recycling plant can reduce environmental impact while improving economic performance for the group of manufacturing facilities.

In DCM Shriram consolidated limited (Kota unit) produces Caustic Soda, calcium Carbide, Cement and PVC Resins.Chlorine and Hydrogen are obtained as by-products from caustic soda production, while Calcium carbide produced is partly sold and partly is treated with water to form Slurry(Aqueous solution of Calcium Hydroxide) and Ethylene. The chlorine and ethylene produced are utilised to form PVC compounds, while the slurry is consumed for Cement production by wet process. Hydrochloric Acid is prepared by direct synthesis where The pure chlorine gas can be combined with hydrogen to produce hydrogen chloride in the presence of UV light.

References

- Lombardi, D. R. and Laybourn, P. (2012), Redefining Industrial Symbiosis. Journal of Industrial Ecology, 16: 28–37. doi: 10.1111/j.1530-9290.2011.00444.x
- Ehrenfeld, J. and Gertler, N. 1997. Industrial Ecology in Practice: The Evolution of Interdependence at Kalundborg, Journal of Industrial Ecology 1(1): 67

- A. H. Nosrat, J. Jeswiet, and J. M. Pearce, "Cleaner Production via Industrial Symbiosis in Glass and Large-Scale Solar Photovoltaic Manufacturing", Science and Technology for Humanity (TIC-STH), 2009 IEEE Toronto International Conference, pp.967-970, 26-27 Sept. 2009

- Costa I., Massard G. and Agarwal A. 2010. Waste management policies for industrial symbiosis development: case studies in European countries, Journal of Cleaner Production 18: 815-822

- M.A. Kreiger, D.R. Shonnard, J.M. Pearce, "Life Cycle Analysis of Silane Recycling in Amorphous Silicon-Based Solar Photovoltaic Manufacturing" Resources, Conservation & Recycling, 70, pp.44-49 (2013)

- Rob Andrews and Joshua Pearce, "Environmental and Economic Assessment of a Greenhouse Waste Heat Exchange", Journal of Cleaner Production 19, pp. 1446-1454 (2011)

Various Concepts Related to Industrial Ecology

The various concepts related to industrial ecology are life-cycle assessment, design for the environment, extended producer responsibility and helix of sustainability. The technique that is used to evaluate environmental impacts that are related with the stages of a product's life is known as life-cycle assessment. This chapter is an overview of the subject matter incorporating all the major aspects of industrial ecology.

Life-cycle Assessment

Life Cycle Assessment Overview

Life-cycle assessment (LCA, also known as life-cycle analysis, ecobalance, and cradle-to-grave analysis) is a technique to assess environmental impacts associated with all the stages of a product's life from raw material extraction through materials processing, manufacture, distribution, use, repair and maintenance, and disposal or recycling. Designers use this process to help critique their products. LCAs can help avoid a narrow outlook on environmental concerns by:

- Compiling an inventory of relevant energy and material inputs and environmental releases;
- Evaluating the potential impacts associated with identified inputs and releases;
- Interpreting the results to help make a more informed decision.

Goals and Purpose

The goal of LCA is to compare the full range of environmental effects assignable to products and services by quantifying all inputs and outputs of material flows and assessing how these material flows affect the environment. This information is used to improve processes, support policy and provide a sound basis for informed decisions.

The term *life cycle* refers to the notion that a fair, holistic assessment requires the assessment of raw-material production, manufacture, distribution, use and disposal including all intervening transportation steps necessary or caused by the product's existence.

There are two main types of LCA. Attributional LCAs seek to establish (or attribute) the burdens associated with the production and use of a product, or with a specific service or process, at a point in time (typically the recent past). Consequential LCAs seek to identify the environmental consequences of a decision or a proposed change in a system under study (oriented to the future), which means that market and economic implications of a decision may have to be taken into account. Social LCA is under development as a different approach to life cycle thinking intended to assess social implications or potential impacts. Social LCA should be considered as an approach that is complementary to environmental LCA.

The procedures of life cycle assessment (LCA) are part of the ISO 14000 environmental management standards: in ISO 14040:2006 and 14044:2006. (ISO 14044 replaced earlier versions of ISO 14041 to ISO 14043.) GHG product life cycle assessments can also comply with specifications such as PAS 2050 and the GHG Protocol Life Cycle Accounting and Reporting Standard.

Four Main Phases

According to the ISO 14040 and 14044 standards, a Life Cycle Assessment is carried out in four distinct phases as illustrated in the figure shown to the right. The phases are often interdependent in that the results of one phase will inform how other phases are completed.

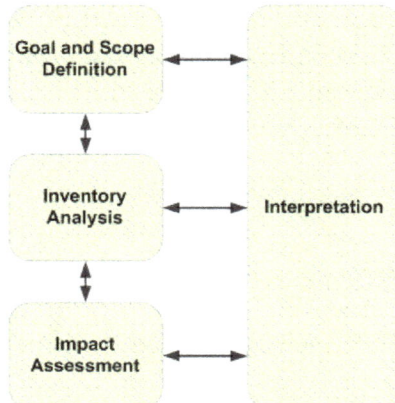

Illustration of LCA phases

Goal and Scope

An LCA starts with an explicit statement of the goal and scope of the study, which sets out the context of the study and explains how and to whom the results are to be communicated. This is a key step and the ISO standards require that the goal and scope of an LCA be clearly defined and consistent with the intended application. The goal and scope document therefore includes technical details that guide subsequent work:

- the functional unit, which defines what precisely is being studied and quantifies the service delivered by the product system, providing a reference to which the inputs and outputs can be related. Further, the functional unit is an important basis that enables alternative goods, or services, to be compared and analyzed. So to explain this a functional system which is inputs, processes and outputs contains a functional unit, that fulfills a function, for example paint is covering a wall, making a functional unit of $1m^2$ covered for 10 years. The functional flow would be the items necessary for that function, so this would be a brush, tin of paint and the paint itself.

- the system boundaries; which are delimitations of which processes that should be included in the analysis of a product system.

- any assumptions and limitations;

- the allocation methods used to partition the environmental load of a process when several products or functions share the same process; allocation is commonly dealt with in one of three ways: system expansion, substitution and partition. Doing this is not easy and different methods may give different results; and

- the impact categories chosen for example human toxicity, smog, global warming, eutrophication.

Life Cycle Inventory

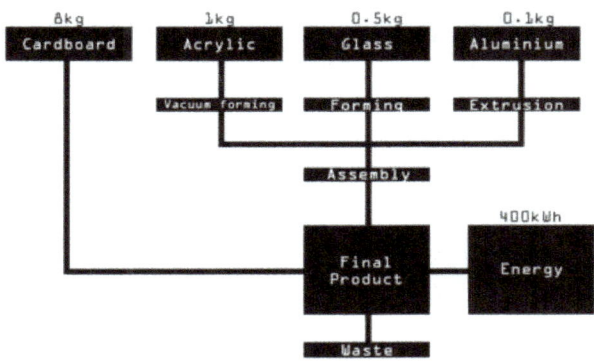

This is an example of a Life-cycle inventory (LCI) diagram

Life Cycle Inventory (LCI) analysis involves creating an inventory of flows from and to nature for a product system. Inventory flows include inputs of water, energy, and raw materials, and releases to air, land, and water. To develop the inventory, a flow model of the technical system is constructed using data on inputs and outputs. The flow model is typically illustrated with a flow chart that includes the activities that are going to be assessed in the relevant supply chain and gives a clear picture of the technical system boundaries. The input and output data needed for the construction of the model are collected for all activities within the system boundary, including from the supply chain (referred to as inputs from the technosphere).

The data must be related to the functional unit defined in the goal and scope definition. Data can be presented in tables and some interpretations can be made already at this stage. The results of the inventory is an LCI which provides information about all inputs and outputs in the form of elementary flow to and from the environment from all the unit processes involved in the study.

Inventory flows can number in the hundreds depending on the system boundary. For product LCAs at either the generic (i.e., representative industry averages) or brand-specific level, that data is typically collected through survey questionnaires. At an industry level, care has to be taken to ensure that questionnaires are completed by a representative sample of producers, leaning toward neither the best nor the worst, and fully representing any regional differences due to energy use, material sourcing or other factors. The questionnaires cover the full range of inputs and outputs, typically aiming to account for 99% of the mass of a product, 99% of the energy used in its production and any environmentally sensitive flows, even if they fall within the 1% level of inputs.

One area where data access is likely to be difficult is flows from the technosphere. The technosphere is more simply defined as the man-made world. Considered by geologists as secondary resources, these resources are in theory 100% recyclable; however, in a practical sense, the primary goal is salvage. For an LCI, these technosphere products (supply chain products) are those that have been produced by man and unfortunately those completing a questionnaire about a process which uses a man-made product as a means to an end will be unable to specify how much of a given input they use. Typically, they will not have access to data concerning inputs and outputs for previous production processes of the product. The entity undertaking the LCA must then turn to secondary sources if it does not already have that data from its own previous studies. National databases or data sets that come with LCA-practitioner tools, or that can be readily accessed, are the usual sources for that information. Care must then be taken to ensure that the secondary data source properly reflects regional or national conditions.

LCI Methods

- Process LCA
- Economic Input Output LCA
- Hybrid Approach

Life Cycle Impact Assessment

Inventory analysis is followed by impact assessment. This phase of LCA is aimed at evaluating the significance of potential environmental impacts based on the LCI flow results. Classical life cycle impact assessment (LCIA) consists of the following mandatory elements:

- selection of impact categories, category indicators, and characterization models;
- the classification stage, where the inventory parameters are sorted and assigned to specific impact categories; and
- impact measurement, where the categorized LCI flows are characterized, using one of many possible LCIA methodologies, into common equivalence units that are then summed to provide an overall impact category total.

In many LCAs, characterization concludes the LCIA analysis; this is also the last compulsory stage according to ISO 14044:2006. However, in addition to the above mandatory LCIA steps, other optional LCIA elements – normalization, grouping, and weighting – may be conducted depending on the goal and scope of the LCA study. In normalization, the results of the impact categories from the study are usually compared with the total impacts in the region of interest, the U.S. for example. Grouping consists of sorting and possibly ranking the impact categories. During weighting, the different environmental impacts are weighted relative to each other so that they can then be summed to get a single number for the total environmental impact. ISO 14044:2006 generally advises against weighting, stating that "weighting, shall not be used in LCA studies intended to be used in comparative assertions intended to be disclosed to the public". This advice is often ignored, resulting in comparisons that can reflect a high degree of subjectivity as a result of weighting.

Interpretation

Life Cycle Interpretation is a systematic technique to identify, quantify, check, and evaluate information from the results of the life cycle inventory and/or the life cycle impact assessment. The results from the inventory analysis and impact assessment are summarized during the interpretation phase. The outcome of the interpretation phase is a set of conclusions and recommendations for the study. According to ISO 14040:2006, the interpretation should include:

- identification of significant issues based on the results of the LCI and LCIA phases of an LCA;
- evaluation of the study considering completeness, sensitivity and consistency checks; and
- conclusions, limitations and recommendations.

A key purpose of performing life cycle interpretation is to determine the level of confidence in the final results and communicate them in a fair, complete, and accurate manner. Interpreting the results of an LCA is not as simple as "3 is better than 2, therefore Alternative A is the best choice"! Interpreting the results of an LCA starts with understanding the accuracy of the results, and ensuring they meet the goal of the study. This is accomplished by identifying the data elements that contribute significantly to each impact category, evaluating the sensitivity of these significant data elements, assessing the completeness and consistency of the study, and drawing conclusions and recommendations based on a clear understanding of how the LCA was conducted and the results were developed.

Reference Test

More specifically, the best alternative is the one that the LCA shows to have the least cradle-to-grave environmental negative impact on land, sea, and air resources.

LCA Uses

Based on a survey of LCA practitioners carried out in 2006 LCA is mostly used to support business strategy (18%) and R&D (18%), as input to product or process design (15%), in education (13%) and for labeling or product declarations (11%). LCA will be continuously integrated into the built environment as tools such as the European ENSLIC Building project guidelines for buildings or developed and implemented, which provide practitioners guidance on methods to implement LCI data into the planning and design process.

Major corporations all over the world are either undertaking LCA in house or commissioning studies, while governments support the development of national databases to support LCA. Of particular note is the growing use of LCA for ISO Type III labels called Environmental Product Declarations, defined as "quantified environmental data for a product with pre-set categories of parameters based on the ISO 14040 series of standards, but not excluding additional environmental information". These third-party certified LCA-based labels provide an increasingly important basis for assessing the relative environmental merits of competing products. Third-party certification plays a major role in today's industry. Independent certification can show a company's dedication to safer and environmental friendlier products to customers and NGOs.

LCA also has major roles in environmental impact assessment, integrated waste management and pollution studies.

Data Analysis

A life cycle analysis is only as valid as its data; therefore, it is crucial that data used for the completion of a life cycle analysis are accurate and current. When comparing different life cycle analyses with one another, it is crucial that equivalent data are available for both products or processes in question. If one product has a much higher availability of data, it cannot be justly compared to another product which has less detailed data.

There are two basic types of LCA data – unit process data and environmental input-output data (EIO), where the latter is based on national economic input-output data. Unit process data are derived from direct surveys of companies or plants producing the product of interest, carried out at a unit process level defined by the system boundaries for the study.

Data validity is an ongoing concern for life cycle analyses. Due to globalization and the rapid pace of research and development, new materials and manufacturing methods are continually being introduced to the market. This makes it both very important and very difficult to use up-to-date information when performing an LCA. If an LCA's conclusions are to be valid, the data must be recent; however, the data-gathering process takes time. If a product and its related processes have not undergone significant revisions since the last LCA data was collected, data validity is not a problem. However, consumer electronics such as cell phones can be redesigned as often as every 9 to 12 months, creating a need for ongoing data collection.

The life cycle considered usually consists of a number of stages including: materials extraction, processing and manufacturing, product use, and product disposal. If the most environmentally harmful of these stages can be determined, then impact on the environment can be efficiently reduced by focusing on making changes for that particular phase. For example, the most energy-intensive life phase of an airplane or car is during use due to fuel consumption. One of the most effective ways to increase fuel efficiency is to decrease vehicle weight, and thus, car and airplane manufacturers can decrease environmental impact in a significant way by replacing heavier materials with lighter ones such as aluminium or carbon fiber-reinforced elements. The reduction during the use phase should be more than enough to balance additional raw material or manufacturing cost.

Data sources are typically large databases, it is not appropriate to compare two options if different data sources have been used to source the data. Data sources include: soca EuGeos' 15804-IA, NEEDS, ecoinvent, PSILCA, ESU World Food, GaBi, ELCD, LC-Inventories.ch, Social Hotspots, ProBas, bioenergiedat, Agribalyse, USDA, Ökobaudat, Agri-footprint

Calculations for impact can then be done by hand, but it is more usual to streamline the process by using software. This can range from a simple spreadsheet, where the user enters the data manually to a fully automated program, where the user is not aware of the source data.

Variants

Cradle-to-grave

Cradle-to-grave is the full Life Cycle Assessment from resource extraction ('cradle') to use phase and disposal phase ('grave'). For example, trees produce paper, which can be recycled into low-energy production cellulose (fiberised paper) insulation, then used as an energy-saving device in the ceiling of a home for 40 years, saving 2,000 times the fossil-fuel energy used in its production. After 40 years the cellulose fibers are replaced and the old fibers are disposed of, possibly incinerated. All inputs and outputs are considered for all the phases of the life cycle.

Cradle-to-gate

Cradle-to-gate is an assessment of a *partial* product life cycle from resource extraction (*cradle*) to the factory gate (i.e., before it is transported to the consumer). The use phase and disposal phase of the product are omitted in this case. Cradle-to-gate assessments are sometimes the basis for environmental product declarations (EPD) termed business-to-business EDPs. One of the significant uses of the cradle-to-gate approach compiles the life cycle inventory (LCI) using cradle-to-gate. This allows the LCA to collect all of the impacts leading up to resources being purchased by the facility. They can then add the steps involved in their transport to plant and manufacture process to more easily produce their own cradle-to-gate values for their products.

Cradle-to-cradle or Closed Loop Production

Cradle-to-cradle is a specific kind of cradle-to-grave assessment, where the end-of-life disposal step for the product is a recycling process. It is a method used to minimize the environmental impact of products by employing sustainable production, operation, and disposal practices and aims to incorporate social responsibility into product development. From the recycling process originate new, identical products (e.g., asphalt pavement from discarded asphalt pavement, glass bottles from collected glass bottles), or different products (e.g., glass wool insulation from collected glass bottles).

Allocation of burden for products in open loop production systems presents considerable challenges for LCA. Various methods, such as the avoided burden approach have been proposed to deal with the issues involved.

Gate-to-gate

Gate-to-gate is a partial LCA looking at only one value-added process in the entire production chain. Gate-to-gate modules may also later be linked in their appropriate production chain to form a complete cradle-to-gate evaluation.

Well-to-wheel

Well-to-wheel is the specific LCA used for transport fuels and vehicles. The analysis is often broken down into stages entitled "well-to-station", or "well-to-tank", and "station-to-wheel" or "tank-to-wheel", or "plug-to-wheel". The first stage, which incorporates the feedstock or fuel production and processing and fuel delivery or energy transmission, and is called the "upstream" stage, while the stage that deals with vehicle operation itself is sometimes called the "downstream" stage. The well-to-wheel analysis is commonly used to assess total energy consumption, or the energy conversion efficiency and emissions impact of marine vessels, aircraft and motor vehicles, including their carbon footprint, and the fuels used in each of these transport modes.

The well-to-wheel variant has a significant input on a model developed by the Argonne National Laboratory. The Greenhouse gases, Regulated Emissions, and Energy use in Transportation (GREET) model was developed to evaluate the impacts of new fuels and vehicle technologies. The model evaluates the impacts of fuel use using a well-to-wheel evaluation while a traditional cradle-to-grave approach is used to determine the impacts from the vehicle itself. The model reports energy use, greenhouse gas emissions, and six additional pollutants: volatile organic compounds (VOCs), carbon monoxide (CO), nitrogen oxide (NOx), particulate matter with size smaller than 10 micrometre (PM10), particulate matter with size smaller than 2.5 micrometre (PM2.5), and sulfur oxides (SOx).

Economic input–output Life Cycle Assessment

Economic input–output LCA (EIOLCA) involves use of aggregate sector-level data on how much environmental impact can be attributed to each sector of the economy and how much each sector purchases from other sectors. Such analysis can account for long chains (for example, building an automobile requires energy, but producing energy requires vehicles, and building those vehicles requires energy, etc.), which somewhat alleviates the scoping problem of process LCA; however, EIOLCA relies on sector-level averages that may or may not be representative of the specific subset of the sector relevant to a particular product and therefore is not suitable for evaluating the environmental impacts of products. Additionally the translation of economic quantities into environmental impacts is not validated.

Ecologically Based LCA

While a conventional LCA uses many of the same approaches and strategies as an Eco-LCA, the latter considers a much broader range of ecological impacts. It was designed to provide a guide to wise management of human activities by understanding the direct and indirect impacts on ecological resources and surrounding ecosystems. Developed by Ohio State University Center for resilience, Eco-LCA is a methodology that quantitatively takes into account regulating and supporting services during the life cycle of economic goods and products. In this approach services are categorized in four main groups: supporting, regulating, provisioning and cultural services.

Exergy Based LCA

Exergy of a system is the maximum useful work possible during a process that brings the system into equilibrium with a heat reservoir. Wall clearly states the relation between exergy analysis and resource accounting. This intuition confirmed by DeWulf and Sciubba lead to Exergo-economic accounting and to methods specifically dedicated to LCA such as Exergetic material input per unit of service (EMIPS). The concept of material input per unit of service (MIPS) is quantified in terms of the second law of thermodynamics, allowing the calculation of both resource input and service output in exergy terms. This exergetic material input per unit of service (EMIPS) has been elaborated for transport technology. The service not only takes into account the total mass to be transported and the total distance, but also the mass per single transport and the delivery time. The applicability of the EMIPS methodology relates specifically to transport system. This model has been further improved by Trancossi who has introduced the friction term, which has not been considered by original EMIPS model, and the key distinction between exergy disruption by payload and by vehicle, focusing on the losses due to vehicle and more effective evaluation of the processes and produced an effective assessment of today transport vehicles. This model is referenced by Indian "Road less traveled" model, which has been developed for minimizing the impact of transports in urban environment.

Life Cycle Energy Analysis

Life cycle energy analysis (LCEA) is an approach in which all energy inputs to a product are accounted for, not only direct energy inputs during manufacture, but also all energy inputs needed to produce components, materials and services needed for the manufacturing process. An earlier term for the approach was *energy analysis*.

With LCEA, the *total life cycle energy input* is established.

Energy Production

It is recognized that much energy is lost in the production of energy commodities themselves, such as nuclear energy, photovoltaic electricity or high-quality petroleum products. *Net energy content* is the energy content of the product minus energy input used during extraction and conversion, directly or indirectly. A controversial early result of LCEA claimed that manufacturing solar cells requires more energy than can be recovered in using the solar cell. The result was refuted. Another new concept that flows from life cycle assessments is Energy Cannibalism. Energy Cannibalism refers to an effect where rapid growth of an entire energy-intensive industry creates a need for energy that uses (or cannibalizes) the energy of existing power plants. Thus during rapid growth the industry as a whole produces no energy because new energy is used to fuel the embodied energy of future power plants. Work has been undertaken in the UK to determine the life cycle energy (alongside full LCA) impacts of a number of renewable technologies.

Energy Recovery

If materials are incinerated during the disposal process, the energy released during burning can be harnessed and used for electricity production. This provides a low-impact energy source, especially when compared with coal and natural gas While incineration produces more greenhouse gas emissions than landfilling, the waste plants are well-fitted with filters to minimize this negative impact. A recent study comparing energy consumption and greenhouse gas emissions from landfilling (without energy recovery) against incineration (with energy recovery) found incineration to be superior in all cases except for when landfill gas is recovered for electricity production.

Criticism

A criticism of LCEA is that it attempts to eliminate monetary cost analysis; that is replace the currency by which economic decisions are made with an energy currency. It has also been argued that energy efficiency is only one consideration in deciding which alternative process to employ, and that it should not be elevated to the only criterion for determining environmental acceptability; for example, simple energy analysis does not take into account the renewability of energy flows or the toxicity of waste products; however the life cycle assessment does help companies become more familiar with environmental properties and improve their environmental system. Incorporating Dynamic LCAs of renewable energy technologies (using sensitivity analyses to project future improvements in renewable systems and their share of the power grid) may help mitigate this criticism.

In recent years, the literature on life cycle assessment of energy technology has begun to reflect the interactions between the current electrical grid and future energy technology. Some papers have focused on energy life cycle, while others have focused on carbon dioxide (CO_2) and other greenhouse gases. The essential critique given by these sources is that when considering energy technology, the growing nature of the power grid must be taken into consideration. If this is not done, a given class of energy technology may emit more CO_2 over its lifetime than it mitigates.

A problem the energy analysis method cannot resolve is that different energy forms (heat, electricity, chemical energy etc.) have different quality and value even in natural sciences, as a consequence of the two main laws of thermodynamics. A thermodynamic measure of the quality of energy is exergy. According to the first law of thermodynamics, all energy inputs should be accounted with equal weight, whereas by the second law diverse energy forms should be accounted by different values.

The conflict is resolved in one of these ways:

- value difference between energy inputs is ignored,
- a value ratio is arbitrarily assigned (e.g., a joule of electricity is 2.6 times more valuable than a joule of heat or fuel input),
- the analysis is supplemented by economic (monetary) cost analysis,
- exergy instead of energy can be the metric used for the life cycle analysis.

Critiques

Life cycle assessment is a powerful tool for analyzing commensurable aspects of quantifiable systems. Not every factor, however, can be reduced to a number and inserted into a model. Rigid system boundaries make accounting for changes in the system difficult. This is sometimes referred to as the boundary critique to systems thinking. The accuracy and availability of data can also contribute to inaccuracy. For instance, data from generic processes may be based on averages, unrepresentative sampling, or outdated results. Additionally, social implications of products are generally lacking in LCAs. Comparative life-cycle analysis is often used to determine a better process or product to use. However, because of aspects like differing system boundaries, different statistical information, different product uses, etc., these studies can easily be swayed in favor of one product or process over another in one study and the opposite in another study based on varying parameters and different available data. There are guidelines to help reduce such conflicts in results but the method still provides a lot of room for the researcher to decide what is important, how the product is typically manufactured, and how it is typically used.

An in-depth review of 13 LCA studies of wood and paper products found a lack of consistency in the methods and assumptions used to track carbon during the product life-cycle. A wide variety of methods and assumptions were used, leading to different and potentially contrary conclusions – particularly with regard to carbon sequestration and methane generation in landfills and with carbon accounting during forest growth and product use.

Streamline LCA

This process includes three steps. First, a proper method should be selected to combine adequate accuracy with acceptable cost burden in order to guide decision making. Actually, in LCA process, besides streamline LCA, Eco-screening and complete LCA are usually considered as well. However, the former one only could provide limited details and the latter one with more detailed information is more expensive. Second, single measure of stress should be selected. Typical LCA output includes resource consumption, energy consumption, water consumption, emission of CO_2, toxic residues and so on. One of these outputs is used as the main factor to measure in streamline LCA. Energy consumption and CO_2 emission are often regarded as "practical indicators". Last, stress selected in step 2 is used as standard to assess phase of life separately and identify the most damaging phase. For instance, for a family car, energy consumption could be used as the single stress factor to assess each phase of life. The result shows that the most energy intensive phase for a family car is the usage stage.

Life Cycle Assessment of Engineered Material in Service plays a significant role in saving energy, conserving resources and saving billions by preventing premature failure of critical engineered component in a machine or equipment. LCA data of surface engineered materials are used to improve life cycle of the engineered component. Life cycle improvement of industrial machineries and equipments including, manufacturing, power generation, transportations, etc. leads to improvement in energy efficiency, sustainability and negating global temperature rise. Estimated reduction in anthropogenic carbon emission is minimum 10% of the global emission.

Design for the Environment

Design for the Environment (DfE) is a design approach to reduce the overall human health and environmental impact of a product, process or service, where impacts are considered across its life cycle. Different software tools have been developed to assist designers in finding optimized products or processes/services. DfE is also the original name of a United States Environmental Protection Agency (EPA) program, created in 1992, that works to prevent pollution, and the risk pollution presents to humans and the environment. This program provides information regarding safer chemical formulations for cleaning and other products. This program was renamed EPA Safer Choice in 2015.

Introduction

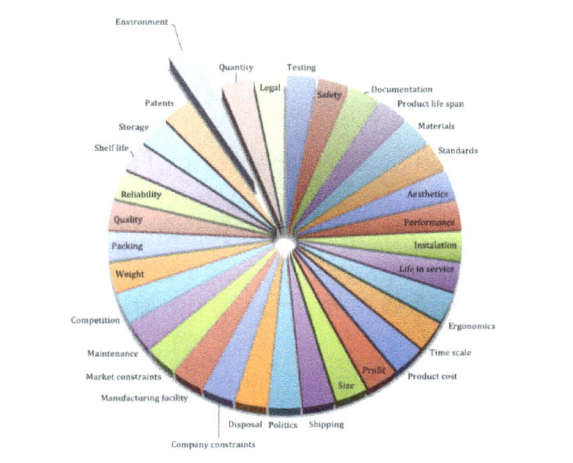

Each piece of the pie chart depicts the role that the individual processes play in the overall design and production of a product. In recent years the environment has begun to play an important role in this diagram. If any piece is missing, production may not be able to occur. This figure was adopted from Conrad Luttropp and Jessica Lagerstedt.

Design for the Environment first guidelines were were written by East Meets West in 1990 in New York, an NGO founded by Anneke van Waesberghe. It became a global movement targeting design initiatives and incorporating environmental motives to improve product design in order to minimize health and environmental impacts by incorporating it from design stage all the way to the manufacturing process. The Design for the Environment (DfE) strategy aims to improve technology and design tactics to expand the scope of products. By incorporating eco-efficiency into design tactics, DfE takes into consideration the entire life-cycle of the product, while still making products usable but minimizing resource use. The key focus of DfE is to minimize the environmental-economic cost to consumers while still focusing on the life-cycle framework of the product. By balancing both customer needs as well as environmental and social impacts DfE aims to "improve the product use experience both for consumers and producers, while minimally impacting the environment".

Design for Environment Practices

Four main concepts that fall under the DfEt umbrella.

- Design for environmental processing and manufacturing: This ensures that raw material extraction (mining, drilling, etc.), processing (processing reusable materials, metal melting, etc.) and manufacturing are done using materials and processes which are not dangerous to the environment or the employees working on said processes. This includes the minimization of waste and hazardous by-products, air pollution, energy expenditure and other factors.

- Design for environmental packaging: This ensures that the materials used in packaging are environmentally friendly, which can be achieved through the reuse of shipping products, elimination of unnecessary paper and packaging products, efficient use of materials and space, use of recycled and/or recyclable materials.

- Design for disposal or reuse: The end-of-life of a product is very important, because some products emit dangerous chemicals into the air, ground and water after they are disposed of in a landfill. Planning for the reuse or refurbishing of a product will change the types of materials that would be used, how they could later be disassembled and reused, and the environmental impacts such materials have.

- Design for energy efficiency: The design of products to reduce overall energy consumption throughout the product's life.

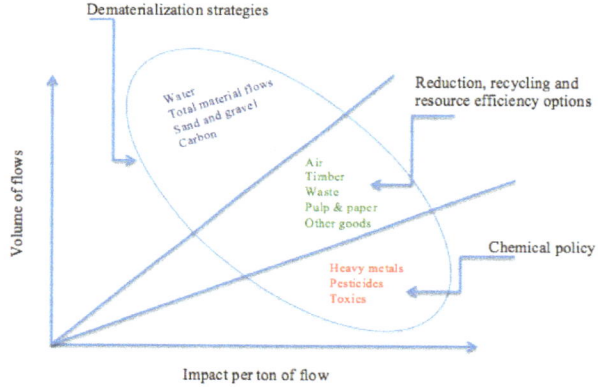

Resource consumption and mitigation strategies for product production which minimizes environmental and health impacts. This figure has been adapted from Spangenberg et al.:

Life cycle assessment (LCA) is employed to forecast the impacts of different (production) alternatives of the product in question, thus being able to choose the most environmentally friendly. A life cycle analysis can serve as a tool when determining the environmental impact of a product or process. Proper LCAs can help a designer compare several different products according to several categories, such as energy use, toxicity, acidification, CO_2 emissions, ozone depletion, resource depletion and many others. By comparing different products, designers can make decisions about which environmental hazard to focus on in order to make the product more environmentally friendly.

Why do Firms Want to Design for the Environment?

Modern day businesses all aim to produce goods at a low cost while maintaining quality, staying competitive in the global marketplace, and meeting consumer preferences for more environmentally friendly products. To help businesses meet these challenges, EPA encourages businesses to incorporate environmental considerations into the design process. The benefits of incorporating DfE include: cost savings, reduced business and environmental risks, expanded business and market opportunities, and to meet environmental regulations.

Companies and Products

Industry	Companies
Technology	IBM, HP, Philips, Sony, Apple, Dell
Food/beverage	Starbucks, Ice Mountain, Coca-Cola, Pepsi
Cleaning	Atlantic Chemical & Equipment Co., American Cleaning Solutions, BCD Supply, Beta Technology, Brighton USA
Automobile/	BMW, GM, Ford

Designfortheenvironment Fig3

- Starbucks: Starbucks is decreasing its carbon footprint by building more energy efficient stores and facilities, conserving energy and water, and purchasing renewable energy credits. Starbucks has achieved LEED certificates in 116 stores in 12 countries. Starbucks has even created a portable, LEED certified store in Denver. It is Starbucks' goal to reduce energy consumption by 25% and to cover 100% of its electricity with renewable energy by 2015.

- Hewlett Packard: HP is working towards reducing energy used in manufacturing, developing materials that have less environmental impact, and designing easily recyclable equipment.

- IBM: Their goal is to extend product life beyond just production, and to use reusable and recyclable products. This means that IBM is currently working on creating products that can be safely disposed of at the end of its product life. They are also reducing consumption of energy to minimize their carbon footprint.

- Philips: For almost 20 years now, sustainable development has been a crucial part of Philips decision making and manufacturing process. Philips' goal is to produce products with their environmental responsibility in mind. Not only are they working on reducing energy during the manufacturing process, Phillips is also participating in a unique project, philanthropy through design. Since 2005, Philips has been working on and developing philanthropy through design. They collaborate with other organizations to use their expertise and innovation to help the more fragile parts of our society.

Besides these large brand names there are several other consumer product companies in the DfE program this including:

- Atlantic Chemical & Equipment Co.
- American Cleaning Solutions
- BCD Supply
- Beta Technology
- Brighton USA

How does a Business Design for the Environment?

A business can design for the environment by:

- Evaluating the human health and environmental impacts of its processes and products.
- Identifying what information is needed to make human health and environment decisions
- Conducting an assessment of alternatives
- Considering cross-media impacts and the benefits of substituting chemicals
- Reducing the use and release of toxic chemicals through the innovation of cleaner technologies that use safer chemicals.
- Implementing pollution prevention, energy efficiency, and other resource conservation measures.
- Making products that can be reused and recycled
- Monitoring the environmental impacts and costs associated with each product or process
- Recognizing that although change can be rapid, in many cases a cycle of evaluation and continuous improvement is needed

Safer Product Labeling Program

The U.S. Environmental Protection Agency DfE labeling program was renamed EPA Safer Choice in 2015.

Current Laws and Regulations Encouraging

The National Ambient Air Quality Standards (NAAQS)

The EPA has imposed the National Ambient Air Quality Standards (NAAQS) to establish an air quality standard across the U.S. The NAAQS sets standards on six main sources of pollutants, which include emissions of: ozone (0.12 ppm per 1 hour), carbon monoxide (35 ppm per 1 hour), pollutant (primary standards), particulate matter (50g/m^3 at an annual arithmetic mean), sulfur dioxide (80g/m^3 at an annual arithmetic mean), nitrogen dioxide (100g/m^3 at an annual arithmetic mean), and lead emissions (1.5g/m^3 at an annual arithmetic mean).

Stratospheric Ozone Protection

The Stratospheric Ozone Protection is under section 602 of the Clean Air Act of 1990. This regulation aims to decrease emission of chlorofluorocarbons (CFCs) and other chemicals that are destroying the stratospheric ozone layer. The protection initiative categorizes ozone-depleting substances into two classes: Class I, and Class II.

Class I substances include 20 different kinds of chemicals and have all been phased-out of production processes since 2000. Class II substances consist of the 33 different HCFCs. The EPA has already begun plans to decrease emissions in HCFCs and plan to completely phase out the class II substances by 2030.

Reporting Requirements for Releases of Toxic Substances

A firm operating in the electronics industry in SIC Codes 20-39 that has more than 10 full-time employees and consumes more than 10,000 lbs per year of any toxic chemical lists in 40 CFR 372.65 must file a toxic release inventory.

Other Regulations

- Hazardous Air Pollutants and Maximum Achievable Control Technology (MACT) Standards
- EPA National Pollutant Discharge Elimination System
- Underground Injection Control Program
- Hazardous Waste Management
- Underground Storage Tank Management

Extended Producer Responsibility

Extended producer responsibility (EPR) is a strategy designed to promote the integration of environmental costs associated with goods throughout their life cycles into the market price of the products. Extended producer responsibility legislation is a driving force behind the adoption of remanufacturing initiatives as it "focuses on the end-of-use treatment of consumer products and has the primary aim to increase the amount and degree of product recovery and to minimize the environmental impact of waste materials"

Tires are an example of products subject to extended producer responsibility in many industrialized countries.

The concept was first formally introduced in Sweden by Thomas Lindhqvist in a 1990 report to the Swedish Ministry of the Environment. In subsequent reports prepared for the Ministry, the following definition emerged: "[EPR] is an environmental protection strategy to reach an environmental objective of a decreased total environmental impact of a product, by making the manufacturer of the product responsible for the entire life-cycle of the product and especially for the take-back, recycling and final disposal.

Definition

EPR uses financial incentives to encourage manufacturers to design environmentally friendly products by holding producers responsible for the costs of managing their products at end of life. This policy approach, which differs from product stewardship, which shares responsibility across the chain of custody of a product, attempts to relieve local governments of the costs of managing certain priority products by requiring manufacturers internalize the cost of recycling within the product price. EPR is based upon the principle that because producers (usually brand owners) have the greatest control over product design and marketing and these same companies have the greatest ability and responsibility to reduce toxicity and waste.

EPR may take the form of a reuse, buy-back, or recycling program. The producer may also choose to delegate this responsibility to a third party, a so-called *producer responsibility organization* (PRO), which is paid by the producer for used-product management. In this way, EPR shifts the responsibility for waste management from government to private industry, obliging producers, importers and/or sellers to internalise waste management costs in their product prices and ensuring the safe handling of their products.

A good example for producer responsibility organizations are the member organizations of PRO EUROPE. PRO EUROPE s.p.r.l. (Packaging Recovery Organisation Europe), founded in 1995, is the umbrella organization for European packaging and packaging waste recovery and recycling schemes. Product stewardship organizations like PRO EUROPE are intended to relieve industrial companies and commercial enterprises of their individual obligation to take back used products through the operation of an organization which fulfills these obligations on a nationwide basis on behalf of their member companies. The aim is to ensure the recovery and recycling of packaging waste in the most economically efficient and ecologically sound manner. In many countries, this is done through suzy the Green Dot (symbol) trademark of which PRO EUROPE is the general licensor. The "Green Dot" has evolved into a proven concept in many countries as implementation of Producer Responsibility. In twenty-five nations companies are now using the "Green Dot" as the financing symbol for the organization of recovery, sorting and recycling of sales packaging.

Take-back

In response to all of the growing problem of excessive waste, several countries adopted waste management policies in which manufacturers are responsible for taking back their products from end users at the end of the products' useful life, or partially financing a collection and recycling infrastructure. These policies were adopted due to the lack of collection infrastructure for certain products that contain hazardous materials, or due to the high costs to local governments of providing such collection services. The primary goals of these take-back laws therefore are to partner with the private sector to ensure that all wastes are managed in a way that protects public health and the environment. The goals of take-back laws are to

1. encourage companies to design products for reuse, recyclability, and materials reduction;
2. correcting market signals to the consumer by incorporating waste management costs into the product's price;
3. promoting innovation in recycling technology.

Take-back programs help promote these goals by creating incentives for companies to redesign their products to minimize waste management costs, by designing their products to contain safer materials (so they do not need to be managed separately) or designing products that are easier to recycle and reuse (so recycling becomes more profitable). The earliest take-back activity began in Europe, where government-sponsored take-back initiatives arose from concerns about scarce landfill space and potentially hazardous substances in component parts. The European Union adopted a directive on Waste Electrical and Electronic Equipment (WEEE).

The purpose of this directive is to prevent the production of waste electronics and also to encourage reuse and recycling of such waste. The directive requires the Member States to encourage design and production methods that take into account the future dismantling and recovery of their products. These take-back programs have been now adopted in nearly every OECD country. In the United States, most of these policies have been implemented at the state level, due to the political impasse at the federal level.

Plastic Bags

United States example:

Recycling, banning, and taxation fails to adequately reduce the pollution caused by plastic bags. An alternative to these policies would be to increase Extended Producer Responsibility. In the US, under the Clinton presidency, the President's Council on Sustainable Development suggested EPR in order to target different participants in the cycle of a product's life. This can however make the product more expensive since the cost must be taken into consideration before being put on the market which is why it is not widely used in the United States currently. Instead there is the banning or taxation of plastic bags that puts the responsibility on the consumers. In the United States EPR has not successfully been made mandatory instead being voluntary. What has been recommended is a comprehensive program which combines taxation, producer responsibility, and recycling to combat pollution.

Electronics

Many governments and companies have adopted Extended Producer Responsibility to help address the growing problem of e-waste—used electronics contain materials that cannot be safely thrown away with regular household trash. In 2007, according to the Environmental Protection Agency, people throw away 2.5 million tons of cell phones, TV's, computers, and printers. Many governments have partnered with corporations in creating the necessary collection and recycling infrastructure. Some argue that local and manufacturer-supported Extended Producer Responsibility laws give manufacturers greater responsibility for the reuse, recycling, and disposal of their own products.

The kinds of chemicals that are found in e-waste that are particularly dangerous to human health and the environment are lead, mercury, brominated flame-retardants, and cadmium. Lead is found in the screens of phones, TV's and computers monitors and can damage kidneys, nerves, blood, bones, reproductive organs, and muscles. Mercury is found in the bulbs in flat screen TV's, laptop screens, and fluorescent bulbs and can cause damage to the kidneys and the nervous system. Brominated flame-retardants are found in cables and plastic cases can cause cancer, disruption of liver function, and nerve damage. Cadmium is found in rechargeable batteries and can cause kidney damage and cancer. Poorer countries are dumping grounds for the United State's e-waste as many governments accept money for disposing this waste on their lands. This dumping of e-waste causes increased health risks for people in these countries, especially ones who work or live close to these dumps.

In the United States, twenty five states have implemented laws that require the recycling of electronic waste. Out of those twenty five states, twenty three have incorporated some form of extended producer responsibility into their laws. According to analysis done by the Product Stewardship Institute, some states have not enacted EPR laws because of a lack of recycling infrastructure and funds for proper e-waste disposal. In contrast, according to a study of EPR legislation done by the Electronics Take-Back Coalition, states that have seen success in their e-waste recycling programs have done so because they have developed a convenient e-waste infrastructure or the state governments have instituted goals for manufacturers to meet. Essentially, these EPR programs have included some driver for increased collection of e-waste and that is why these states have seen a greater impact on proper e-waste disposal than others. Additionally, advocates for EPR argue that including "high expectations for performance" into the laws and ensuring that those are only minimum requirements, contribute to making the laws successful. In this way, manufacturers can be incentivized to collect more and dispose of e-waste more properly. Finally, the larger the scope of products that can be collected, the more e-waste will be disposed of properly.

Similar laws have been passed in other parts of the world as well. The European Union has taken steps in order to combat the issue of electronic waste management. They have restricted the use of harmful substances in member countries and have made it illegal to export waste. The Chinese laws regarding e-waste are similar to the ones in the EU, but they focus on banning the import of e-waste. This has proven to be difficult, however, because illegal smuggling of waste still occurs in the country. In order to dispose of e-waste in China today, a license is required and plants are held responsible for treating pollution.

Advantages

When producers either face a financial or physical burden of recycling their electronics after use, they may be incentivized to design more sustainable, less toxic, and easily recyclable electronics. Using fewer materials and designing products to last longer can directly reduce producers' end-of-life costs. Thus, Extended Producer Responsibility is often cited as one way to fight planned obsolescence, because it financially encourages manufacturers to design for recycling and make products last longer.

Disadvantages

Some people have concerns about extended producer responsibility programs for complex electronics that can be difficult to safely recycle, such as Lithium-ion polymer batteries. Others worry such laws could increase the cost of electronics because producers would be adding recycling costs into the initial price tag. When companies are required to transport their products to a recycling facility, it can be expensive if the product contains hazardous materials and does not have a scrap value, such as with CRT televisions, which can contain up to 5 pounds of lead. Organizations and researchers against EPR claim that the mandate would slow technical innovation and impede technological process. Other critics are concerned that manufacturers may use takeback programs to take secondhand electronics off the reuse market, by shredding rather than reusing or repairing goods that come in for recycling. In addition, another argument against EPR is that EPR policies are not accelerating environmentally-friendly designs because "manufacturers are already starting to moving toward reduced material-use per unit of output, reduced energy use in making and delivering each product, and improved environmental performance." The Reason Foundation argues that EPR is not clear in the way fees are established for the particular recycling processes. Fees are set in place to help incentivize recycling but that may deter the use of manufacturing with better materials for the different electronic products. There are not set fees for certain materials so confusion occurs when companies do not know what design features to include in their devices.

Implementation

EPR has been implemented in many forms, which may be classified into three major approaches:

- Mandatory
- Negotiated
- Voluntary

It is perhaps because of the tendency of economic policy in market-driven economies not to interfere with consumers' preferences that the producer-centric representation is the dominant form of viewing the environmental impacts of industrial production: in statistics on energy, emissions, water, etc., impacts are almost always presented as attributes of industries ('on-site' or 'direct' allocation) rather than as attributes of the supply chains of products for consumers. On a smaller scale, most existing schemes for corporate sustainability reporting include only impacts that arise out of operations controlled by the reporting company, and not supply-chain impacts According to this world view, "upstream and downstream [environmental] impacts are allocated to their immediate producers. The institutional setting and the different actors' spheres of influence are not reflected".

On the other hand, a number of studies have highlighted that final consumption and affluence, especially in the industrialised world, are the main drivers for the level and growth of environmental pressure. Even though these studies provide a clear incentive for complementing producer-focused environmental policy with some consideration for consumption-related aspects, demand-side measures to environmental problems are rarely exploited.

The nexus created by the different views on impacts caused by industrial production is exemplified by several contributions to the discussion about producer or consumer responsibility for greenhouse gas emissions. Emissions data are reported to the IPCC as contributions of producing industries located in a particular country rather than as embodiments in products consumed by a particular population, irrespective of productive origin. However, especially for open economies, taking into account the greenhouse gases embodied in internationally traded commodities can have a considerable influence on national greenhouse gas balance sheets. Assuming consumer responsibility, exports have to be subtracted from, and imports added to national greenhouse gas inventories. In Denmark for example, Munksgaard and Pedersen (2001) report that a significant amount of power and other energy-intensive commodities are traded across Danish borders, and that between 1966 and 1994 the Danish foreign trade balance in terms of CO_2 developed from a 7 Mt deficit to a 7 Mt surplus, compared to total emissions of approximately 60 Mt. In particular, electricity traded between Norway, Sweden and Denmark is subject to large annual fluctuations due to varying rainfall in Norway and Sweden. In wet years Denmark imports hydro-electricity whereas electricity from coal-fired power plants is exported in dry years. The official Danish emissions inventory includes a correction for electricity trade and thus applies the consumer responsibility principle.

Similarly, at the company level, "when adopting the concept of eco-efficiency and the scope of an environmental management system stated in for example ISO 14001, it is insufficient to merely report on the carbon dioxide emissions limited to the judicial borders of the company". 7 "Companies must recognise their wider responsibility and manage the entire life-cycle of their products Insisting on high environmental standards from suppliers and ensuring that raw materials are extracted or produced in an environmentally conscious way provides a start." A life-cycle perspective is also taken in Extended Producer Responsibility (EPR) frameworks: "Producers of products should bear a significant degree of responsibility (physical and/or financial) not only for the environmental impacts of their products downstream from the treatment and disposal of their product, but also for their upstream activities inherent in the selection of materials and in the design of products." "The major impetus for EPR came from northern European countries in the late 1980s and early 1990s, as they were facing severe landfill shortages. As a result, EPR is generally applied to post-consumer wastes which place increasing physical and financial demands on municipal waste management."

EPR has rarely been consistently quantified. Moreover, applying conventional life cycle assessment, and assigning environmental impacts to producers and consumers can lead to double-counting. Using input-output analysis, researchers have attempted for decades to account for both producers and consumers in an economy in a consistent way. Gallego and Lenzen demonstrate and discuss a method of consistently delineating producers' supply chains, into mutually exclusive and collectively exhaustive responsibilities to be shared by all agents in an economy. Their method is an approach to allocating responsibility across agents in a fully inter-connected circular system. Upstream and downstream environmental impacts are shared between all agents of a supply chain – producers and consumers.

Examples

Auto Recycling Nederland (ARN) is a Producer Responsibility Organisation (PRO) that organises vehicle recycling in the Netherlands. An advanced recycling fee is charged to those who purchase a new vehicle and is used to fund the recycling of it at the end of its useful life. The PRO was set up to satisfy the European Union End of Life Vehicles Directive.

The Swiss Association for Information, Communication and Organisational Technology (SWICO), an ICT industry organisation, became a PRO to address the problem of electronic waste.

Results

In Germany, since the adoption of EPR, "between 1991 and 1998, the per capita consumption of packaging was reduced from 94.7 kg to 82 kg, resulting in a reduction of 13.4%". Furthermore, due to Germany's influence in EPR, the "European Commission developed one waste directive" for all of member states (Hanisch 2000). One major goal was to have all member states recycle "25% of all packaging material" and have accomplished the goal.

In the United States, EPR is gaining popularity "with 40 such laws enacted since 2008. In 2010 alone, 38 such EPR bills were introduced in state legislatures across the United States, and 12 were signed into law." However, these laws are only at the state level as there are no federal laws for EPR. So far, "only a handful of states have imposed five to six EPR laws as well as 32 states having at least one EPR law".

Helix of Sustainability

The Helix of Sustainability

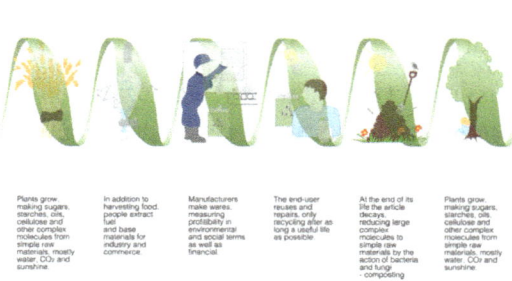

The helix of sustainability - the Carbon cycle ideal for manufacture and use

The international recycling symbol - not nature identical.

The helix of sustainability is a concept coined to help manufacturing industry move to more sustainable practices by mapping its models of raw material use and reuse onto those of nature. The environmental benefits of the use crop origin sustainable materials have been assumed to be self-evident, but as the debate on food vs fuel shows, the whole product life cycle must be examined in the light of social and environmental effects in addition to technical suitability and profitability.

The helix of sustainability is a concept created as a representation of the total systems approach to gain full advantage from manufacturing with sustainable materials, particularly biopolymers and biocomposites. In 2004 the concept was presented by Professor John Wood, then Chair of the Materials Foresight Panel at a DTI event hosted by the then Secretary of State for Industry (Jacqui Smith). In the same year it was also used in the European Science Foundation exploratory workshop on environmentally friendly composites.

The advantages of working with crop origin raw materials are readily observed if the social and environmental impacts are considered as well as monetary cost (the Triple bottom line), and the helix of sustainability helps to demonstrate this. For the full potential of biopolymers to be realised it is essential that attention is paid to every aspect of the manufacturing process from design (how to cope with the uncertainties in properties associated with crop origin materials?), manufacture (can existing technologies be used?), through to end-of-life (can the redundant article be fed back into the materials cycle?). The entire supply chain must be considered because decisions taken at the design stage have significant effects right through the life of an article. Low cost assembly techniques (e.g. snap-fits) may make dismantling or repair uneconomical. However, if say, an easy-to-dismantle car is built, will there be any effect on the ability of the vehicle to absorb energy in a crash? At an even more fundamental level, what will be the social and environmental of the change in crop growing patterns. This low environmental impact approach to manufacturing is seen as an extension of waste reduction techniques such as lean manufacturing.

Conventional cycles of use and reuse are circular. Consider the mechanical recovery of conventional polymers. A complex infrastructure is needed to recover the material at the end of an article's useful life. At the end of an article's life - say a PET carbonated drink bottle, the article must be separated from the waste stream, either by the consumer who throws it away, or by manual labour at the rubbish dump. It must then be transported to some facility to be reprocessed (using more labour and energy) back into a raw material. The heat and shear forces associated with the process of remanufacture tends to produce material with slightly degraded properties compared to the original material.

For sustainable material articles there is not such a great requirement for a dedicated recovery infrastructure. If a litter lout throws a crop origin biodegradable article on the ground, it will ultimately biodegrade into humus, water, and non-fossil CO_2. If the article is placed into a compostable waste stream, the humus can then be used as fertiliser for the next generation of crops, there is also no requirement to sort biopolymer articles as there is with fossil polymer recycling. Note difference between landfill and compost - the limited biological activity in landfill is slow, and mostly anaerobic resulting in the production of methane, whereas composting is a rapid aerobic process resulting in humus, water and non-fossil CO_2. The energy bill for breaking down biodegradables into the fundamental building block molecules, and then reassembling them into usable raw materials is large, but is uses direct solar energy rather than metered electricity. There is also no loss of properties with successive journeys through the cycle.

References

- Finnveden, G., Hauschild, M.Z., Ekvall, T., Guinée, J., Heijungs, R., Hellweq, S., Koehler, A., Pennington, D. & Suh, S. (2009). Recent developments in Life Cycle Assessment. Journal of Environmental Management 91(1), 1-21

- "PAS 2050:2011 Specification for the assessment of the life cycle greenhouse gas emissions of goods and services". BSI. Retrieved on: 25 April 2013

- Cooper, J.S.; Fava, J. (2006). "Life Cycle Assessment Practitioner Survey: Summary of Results". Journal of Industrial Ecology

- Dewulf, J.; Van Langenhove, H.; Muys, B.; Bruers, S.; Bakshi, B. R.; Grubb, G. F.; Sciubba, E. (2008). "Exergy: its potential and limitations in environmental science and technology" (PDF). Environmental Science & Technology. 42 (7): 2221–2232. doi:10.1021/es071719a

- Hendrickson, C. T., Lave, L. B., and Matthews, H. S. (2005). Environmental Life Cycle Assessment of Goods and Services: An Input–Output Approach, Resources for the Future Press ISBN 1-933115-24-6

- Jiménez-González, C.; Kim, S.; Overcash, M. Methodology for developing gate-to-gate Life cycle inventory information. The International Journal of Life Cycle Assessment 2000, 5, 153–159.

- R. Kenny; C. Law; J.M. Pearce (2010). "Towards Real Energy Economics: Energy Policy Driven by Life-Cycle Carbon Emission". Energy Policy. 38 (4): 1969–1978. doi:10.1016/j.enpol.2009.11.078

- Wall, G., & Gong, M. (2001). On exergy and sustainable development—Part 1: Conditions and concepts. Exergy, An International Journal, 1(3), 128-145

- Brinkman, Norman; Eberle, Ulrich; Formanski, Volker; Grebe, Uwe-Dieter; Matthe, Roland (15 April 2012). "Vehicle Electrification - Quo Vadis". VDI. Retrieved 27 April 2013

- Sciubba, E (2004). "From Engineering Economics to Extended Exergy Accounting: A Possible Path from Monetary to Resource-Based Costing" (PDF). Journal of Industrial Ecology. 8 (4): 19–40. doi:10.1162/1088198043630397

- McManus, M (2010). "Life cycle impacts of waste wood biomass heating systems: A case study of three UK based systems". Energy. 35 (10): 4064–4070. doi:10.1016/j.energy.2010.06.014

- Hammond, Geoffrey P. (2004). "Engineering sustainability: thermodynamics, energy systems, and the environment" (PDF). International Journal of Energy Research. 28 (7): 613–639. doi:10.1002/er.988

- Curran, Mary Ann. "Life Cycle Analysis: Principles and Practice" (PDF). Scientific Applications International Corporation. Archived from the original (PDF) on 18 October 2011. Retrieved 24 October 2011

- Pehnt, Martin (2006). "Dynamic life cycle assessment (LCA) of renewable energy technologies". Renewable Energy: an International Journal. 31 (1): 55–71. doi:10.1016/j.renene.2005.03.002

- S. Singh; B. R. Bakshi (2009). "Eco-LCA: A Tool for Quantifying the Role of Ecological Resources in LCA". International Symposium on Sustainable Systems and Technology: 1–6. ISBN 978-1-4244-4324-6. doi:10.1109/ISSST.2009.5156770

- Luttropp, Conrad; Jessica Lagerstedt (2006). "EcoDesign and The Ten Golden Rules: generic advice for merging environmental aspects into product development". Journal of Cleaner Production

- Franklin Associates, A Division of Eastern Research Group. "Cradle-to-gate Life Cycle Inventory of Nine Plastic Resins and Four Polyurethane Precursors" (PDF). The Plastics Division of the American Chemistry Council. Retrieved 31 October 2012

- Nash, Jennifer, and Christopher Bosso. "Extended Producer Responsibility in the United States." Journal of Industrial Ecology 17.2 (2013): 175-85

- Allen, S.R., G.P. Hammond, H. Harajli, C.I. Jones, M.C. McManus and A.B. Winnett (2008). "Integrated appraisal of micro-generators: methods and applications". 161 (2): 73–86. doi:10.1680/ener.2008.1+61.2.73

Industrial Ecology: Innovative Approaches

The application of mechatronical technology to reduce the cost of ownership of machines and the ecological impact they have is known as ecomechatronics. Cradle-to-cradle design, integrated chain management, environmental management system, ecodesign, ecological modernization and regenerative design are some significant and important topics related to industrial ecology. The following chapter unfolds its crucial aspects in a critical yet systematic manner.

Ecomechatronics

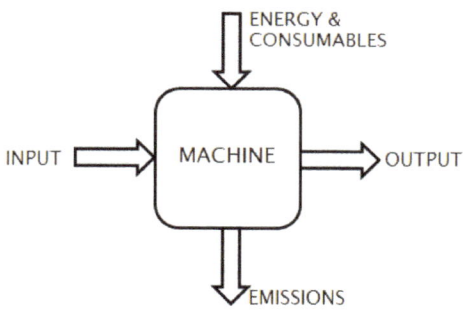

Machine as a system requiring energy and consumables to transform an input into an output, thereby generating emissions (heat, noise)

Ecomechatronics is an engineering approach to developing and applying mechatronical technology in order to reduce the ecological impact and total cost of ownership of machines. It builds upon the integrative approach of mechatronics, but not with the aim of only improving the functionality of a machine. Mechatronics is the multidisciplinary field of science and engineering that merges mechanics, electronics, control theory, and computer science to improve and optimize product design and manufacturing. In ecomechatronics, additionally, functionality should go hand in hand with an efficient use and limited impact on resources. Machine improvements are targeted in 3 key areas: energy efficiency, performance and user comfort (noise & vibrations).

Description

Among policy makers and manufacturing industries there is a growing awareness of the scarcity of resources and the need for sustainable development. This results in new regulations with respect to the design of machines (e.g. European Ecodesign Directive 2009/125/EC) and to a paradigm shift in the global machines market: "instead of maximum profit from minimum capital, maximum added value must be generated from minimal resources". Manufacturing industries increasingly require high performance machines that use resources (energy, consumables) economically in a human-centered production. Machine building companies and original equipment manufacturers are thus urged to respond to this market demand with a new generation of high performance machines with higher energy efficiency and user comfort.

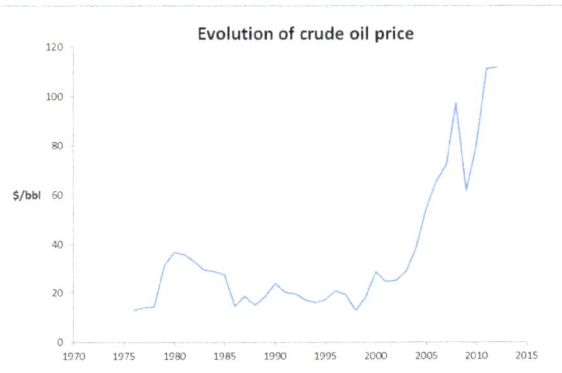

Evolution of crude oil price. Data source: Statistical Review of World Energy 2013, BP

A reduction of the energy consumption lowers energy costs and reduces environmental impact. Typically more than 80% of the total-life-cycle impact of a machine is attributed to its energy consumption during the use phase. Therefore, improving a machine's energy efficiency is the most effective way of reducing its environmental impact. Performance quantifies how well a machine executes its function and is typically related to productivity, precision and availability. User comfort is related to the exposure of operators and the environment to noise & vibrations due to machine operation.

Since energy efficiency, performance and noise & vibrations are coupled in a machine they need to be addressed in an integrated way in the design phase. Example of the interrelation between the 3 key areas: with increasing machine speed typically the machine's productivity increases, but energy consumption will increase as well and machine vibrations may grow such that machine accuracy (e.g. positioning accuracy) and availability (due to downtime and maintenance) decrease. Ecomechatronical design deals with the trade-off between these key areas.

Approach

Ecomechatronics impacts the way mechatronical systems and machines are being designed and implemented. Therefore, the transformation to a new generation of machines concerns knowledge institutes, original equipment manufacturers, CAE software suppliers, machine builders and industrial machine owners. The fact that about

80% of the environmental impact of a machine is determined by its design puts emphasis on making the right technological design choices. A model-based, multidisciplinary design approach is required in order to address the energy efficiency, performance and user comfort of a machine in an integrated way.

The key enabling technologies can be categorized in machine components, machine design methods & tools, and machine control. A few examples are listed below per category.

Machine Components

- Energy efficient electrical motors: cf. energy efficiency classes of electric motors, ecodesign requirements for electric motors
- Variable frequency drives: variable motor speed enables energy reduction with respect to fixed speed applications
- Variable hydraulic pumps: energy reduction by adapting to required pressure and flow (e.g. variable displacement pump, load sensing pump)
- Energy storage technologies: electrical (battery, capacitor, supercapacitor), hydraulical (accumulator), kinetic energy (flywheel), pneumatic, magnetic (superconducting magnetic energy storage)

Design Methods & Tools

- Energetic simulations: using energetic machine models and empirical data (e.g. energy efficiency maps) to estimate the machine's energy consumption in the design phase
- Energy demand optimization: e.g. load leveling in order to avoid peaks in power demand
- Hybridization: applying at least one other, intermediate energy form in order to reduce primary power source consumption e.g. in vehicles with internal combustion engines
- Vibro-acoustic analysis: study of the noise & vibrations signature of a machine in order to localize and differentiate between their root causes
- Multibody modeling: simulation of the interaction forces and displacements of coupled rigid bodies, e.g. to assess the effect of vibration dampers on a mechanical structure
- Active vibration damping: e.g. use of piezoelectric bearings for active control of machine vibrations

- Rapid control prototyping: provides a fast and inexpensive way for control and signal processing engineers to verify designs early and evaluate design tradeoffs

Machine Control

- Energy consumption minimization: control signals are optimized for minimum energy consumption

- Energy management of energy storage systems: controlling the power flows and state-of-charge of an energy storage system with the aim of achieving maximum energy benefit and maximum system lifespan

- Model-based control: taking advantage of system models to improve the outcome (accuracy, reaction time,) of the controlled system

- (Self-)learning control: control self-adapting to the system and its changing environment, reducing the need for control parameter tuning and adaptation by the control engineer

- Optimal machine control: the control of the system is regarded as an optimization problem to which the control rules are considered the optimal solution

Applications

Some examples of ecomechatronical system applications are:

- Komatsu PC200-8 Hybrid: the world's first hybrid excavator has an energy storage system based on supercapacitors. The energy recuperation in the hydraulic drive line during braking results in a significant improvement of fuel economy.

- Hybrid bus: different hybrid bus types have been commercialized (e.g. ExquiCity bus by Van Hool), using fuel cells or a diesel engine as a primary energy source and batteries and/or supercapacitors as energy storage systems.

- Hybrid tram vehicle: hybridization in tram vehicles enables energy recuperation as well as mobility without overhead lines, as applied in e.g. some of the Combino Supra tram vehicles by Siemens Transportation Systems. The system uses a combination of traction batteries and supercapacitors.

Cradle-to-cradle Design

Cradle-to-cradle design (also referred to as Cradle to Cradle, C2C, cradle 2 cradle, or regenerative design) is a biomimetic approach to the design of products and systems that models human industry on nature's processes viewing materials as nutrients circulating in healthy, safe metabolisms. The term itself is a play on the popular corporate phrase "Cradle to Grave," implying that the C2C model is sustainable and considerate of life and future generations (i.e. from the birth, or "cradle," of one generation to the next versus from birth to death, or "grave," within the same generation.)

C2C suggests that industry must protect and enrich ecosystems and nature's biological metabolism while also maintaining a safe, productive technical metabolism for the high-quality use and circulation of organic and technical nutrients. It is a holistic economic, industrial and social framework that seeks to create systems that are not only efficient but also essentially waste free. The model in its broadest sense is not limited to industrial design and manufacturing; it can be applied to many aspects of human civilization such as urban environments, buildings, economics and social systems.

The term Cradle to Cradle is a registered trademark of McDonough Braungart Design Chemistry (MBDC) consultants. Cradle to Cradle product certification began as a proprietary system; however, in 2012 MBDC turned the certification over to an independent non-profit called the Cradle to Cradle Products Innovation Institute. Independence, openness, and transparency are the Institute's first objectives for the certification protocols. The phrase "cradle to cradle" itself was coined by Walter R. Stahel in the 1970s. The current model is based on a system of "lifecycle development" initiated by Michael Braungart and colleagues at the *Environmental Protection Encouragement Agency* (EPEA) in the 1990s and explored through the publication *A Technical Framework for Life-Cycle Assessment*.

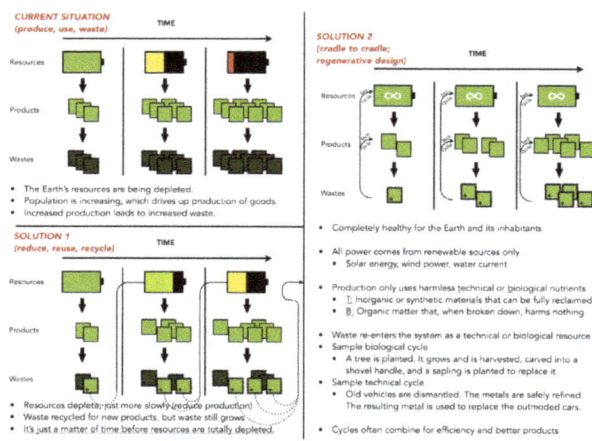

The current economic system, the current solution (the 3Rs), and the C2C framework as an alternative solution.

In 2002, Braungart and William McDonough published a book called *Cradle to Cradle: Remaking the Way We Make Things*, a manifesto for cradle to cradle design that gives specific details of how to achieve the model. The model has been implemented by a number of companies, organizations and governments around the world, predominantly in

Industrial Ecology: Innovative Approaches

the European Union, China and the United States. Cradle to cradle has also been the subject of many documentary films, including the critically acclaimed *Waste=Food*.

Introduction

In the cradle to cradle model, all materials used in industrial or commercial processes—such as metals, fibers, dyes—fall into one of two categories: "technical" or "biological" nutrients. *Technical nutrients* are strictly limited to non-toxic, non-harmful synthetic materials that have no negative effects on the natural environment; they can be used in continuous cycles as the same product without losing their integrity or quality. In this manner these materials can be used over and over again instead of being "downcycled" into lesser products, ultimately becoming waste.

Biological Nutrients are organic materials that, once used, can be disposed of in any natural environment and decompose into the soil, providing food for small life forms without affecting the natural environment. This is dependent on the ecology of the region; for example, organic material from one country or landmass may be harmful to the ecology of another country or landmass.

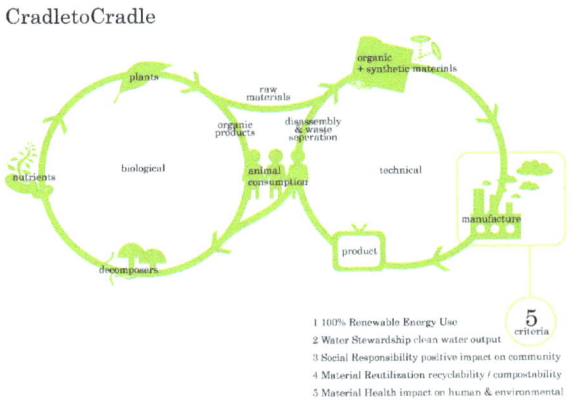

Biological and Technical Cycles

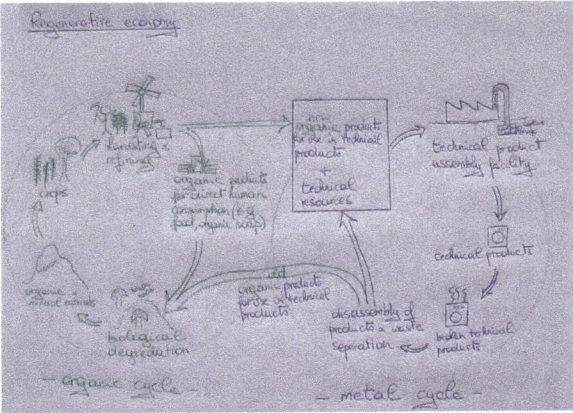

Biological and technical cycle

The two types of materials each follow their own cycle in the regenerative economy envisioned by Keunen and Huizing.

Structure

Initially defined by McDonough and Braungart, the Cradle to Cradle Products Innovation Institute's five certification criteria are:

- Material health, which involves identifying the chemical composition of the materials that make up the product. Particularly hazardous materials (e.g. heavy metals, pigments, halogen compounds etc.) have to be reported whatever the concentration, and other materials reported where they exceed 100 ppm. For wood, the forest source is required. The risk for each material is assessed against criteria and eventually ranked on a scale with green being materials of low risk, yellow being those with moderate risk but are acceptable to continue to use, and red for materials that have high risk and need to be phased out. Grey for materials with incomplete data. The method uses the term 'risk' in the sense of hazard (as opposed to consequence and likelihood).

- Material reutilization, which is about recovery and recycling at the end of product life.

- Assessment of energy required for production, which for the highest level of certification needs to be based on at least 50% renewable energy for all parts and subassemblies.

- Water, particularly usage and discharge quality.

- Social responsibility, which assesses fair labor practices.

The certification is available at several levels: basic, silver, gold, platinum, with more stringent requirements at each. Prior to 2012, MBDC controlled the certification protocol.

Health

Currently, many human beings come into contact or consume, directly or indirectly, many harmful materials and chemicals daily. In addition, countless other forms of plant and animal life are also exposed. C2C seeks to remove dangerous *technical nutrients* (synthetic materials such as mutagenic materials, heavy metals and other dangerous chemicals) from current life cycles. If the materials we come into contact with and are exposed to on a daily basis are not toxic and do not have long term health effects, then the health of the overall system can be better maintained. For example, a fabric factory can eliminate all harmful *technical nutrients* by carefully reconsidering what chemicals they use in their dyes to achieve the colours they need and attempt to do so with fewer base chemicals.

Economics

The use of a C2C model often lowers the financial cost of systems. For example, in the redesign of the Ford River Rouge Complex, the planting of Sedum (stonecrop) vegetation on assembly plant roofs retains and cleanses rain water. It also moderates the internal temperature of the building in order to save energy. The roof is part of an $18 million rainwater treatment system designed to clean 20 billion US gallons (76,000,000 m³) of rainwater annually. This saved Ford $50 million that would otherwise have been spent on mechanical treatment facilities. If products are designed according to C2C design principles, they can be manufactured and sold for less than alternative designs. They eliminate the need for waste disposal such as landfills.

Definitions

- Cradle to Cradle a play on the phrase "Cradle to Grave", implying that the C2C model is sustainable and considerate of life and future generations.
- Technical nutrients are basically inorganic or synthetic materials manufactured by humans—such as plastics and metals—that can be used many times over without any loss in quality, staying in a continuous cycle.
- Biological nutrients and materials are organic materials that can decompose into the natural environment, soil, water, etc. without affecting it in a negative way, providing food for bacteria and microbiological life.
- Materials are usually referred to as the building blocks of other materials, such as the dyes used in colouring fibers or rubbers used in the sole of a shoe.
- Downcycling is the reuse of materials into lesser products. For example, a plastic computer case could be downcycled into a plastic cup, which then becomes a park bench, etc.; this may eventually lead to waste. In conventional understanding, this is no different from recycling that produces a supply of the same product or material.
- Waste = Food is a basic concept of organic waste materials becoming food for bugs, insects and other small forms of life who can feed on it, decompose it and return it to the natural environment which we then indirectly use for food ourselves.

Existing Synthetic Materials

The question of how to deal with the countless existing *technical nutrients* (synthetic materials) that cannot be recycled or reintroduced to the natural environment is dealt with in C2C design. The materials that can be reused and retain their quality can be used within the technical nutrient cycles while other materials are far more difficult to deal with, such as plastics in the Pacific Ocean.

Hypothetical Examples

One effective example is a shoe that is designed and mass-produced using the C2C model. The sole might be made of "biological nutrients" while the upper parts might be made of "technical nutrients." The shoe is mass-produced at a manufacturing plant that utilises its waste material by putting it back into the cycle; an example of this is using off-cuts from the rubber soles to make more soles instead of merely disposing of them (this is dependent on the technical materials not losing their quality as they are reused). Once the shoes have been manufactured, they are distributed to retail outlets where the customer buys the shoe at a fraction of the price they would normally pay for a shoe of comparable aspects; the customer is only paying for the use of the materials in the shoe for the period of time that they will be using the shoe. When they outgrow the shoe or it is damaged, they return it to the manufacturer. When the manufacturer separates the sole from the upper parts (separating the technical and biological nutrients), the biological nutrients are returned to the natural environment while the technical nutrients are used to create the sole of another shoe.

Another example of C2C design is a disposable cup, bottle, or wrapper made entirely out of biological materials. When the user is finished with the item, it can be disposed of and returned to the natural environment; the cost of disposal of waste such as landfill and recycling is eliminated. The user could also potentially return the item for a refund so it can be used again.

Ford Model U is a design concept of a car, made completely from cradle-to-cradle materials. It also uses hydrogen propulsion.

Finished Products

- Cradle-to-cradle shoes have been made through the Nike Considered project.
- The Edag light car
- Rohner Textile AG Climatex-textile
- Biofoam; a cradle-to-cradle alternative to expanded polystyrene
- Sewage sludge processing plants are facilities that create fertiliser from sewage sludge. This approach is green retrofit for the current (inefficient) system of organic waste disposal; as composting toilets are a better approach in the long run.
- Aquion Energy large scale batterys
- Ecovative Design packaging and insulation made from waste by binding it together with Mycelium

Implementation

The C2C model can be applied to almost any system in modern society: urban environments, buildings, manufacturing, social systems. 5 steps are outlined in *Cradle to Cradle – Remaking the way we make things*:

- Get "free of" known culprits
- Follow informed personal preferences
- Create "passive positive" lists – lists of materials used categorised according to their safety level
- The X List – substances that must be phased out, such as teratogenic, mutagenic, carcinogenic.
- The Gray List – problematic substances that are not so urgently in need of phasing out
- The P List – the "positive" list, substances actively defined as safe for use
- Activate the positive list
- Reinvent – the redesign of the former system

Products that adhere to all steps can generally be granted a certification. Two certifications used for cradle-to-cradle products include Leadership in Energy and Environmental Design (LEED) and BRE Environmental Assessment Method (BREEAM).

C2C principles were first applied to systems in the early 1990s by Braungart's Hamburger Umweltinstitut (HUI) and The Environmental Institute in Brazil for biomass nutrient recycling of effluent to produce agricultural products and clean water as a by-product.

In 2005, William McDonough helped found the Center for Eco-Intelligent Management at Instituto de Empresa Business School. The center's research produced the Biosphere Rules, a set of five implementation principles that facilitate the adoption of closed loop production approaches with a minimum of disruption for established companies.

In 2007, MBDC and the EPEA formed a strategic partnership with global materials consultancy Material ConneXion to help promote and disseminate C2C design principles by providing greater global access to C2C material information, certification and product development.

As of January 2008, Material ConneXion's Materials Libraries in New York, Milan, Cologne, Bangkok and Daegu, Korea started to feature C2C assessed and certified materials and, in collaboration with MBDC and EPEA, the company now offers C2C Certification, and C2C product development.

While the C2C model has influenced the construction or redevelopment of many smaller buildings, several large companies, organisations and governments have also implemented the C2C model and its ideas and concepts:

Major Implementations

- The Lyle Center for Regenerative Studies incorporates cradle to cradle systems throughout the center. The use of the term C2C is replaced with Regenerative.

- The Chinese Government is constructing many cities like Huangbaiyu based on C2C principles, utilising the rooftops for agriculture.

- The Ford River Rouge Complex redevelopment. Cleaning 20 billion US gallons (76,000,000 m^3) of rainwater annually.

- The Netherlands Institute of Ecology (NIOO-KNAW) will make its laboratory and office complex completely cradle to cradle compliant

- Several private houses and communal buildings in the Netherlands

- Fashion Positive, an initiative to assist the fashion world in implementing the cradle-to-cradle model in five areas: material health, material reuse, renewable energy, water stewardship and social fairness.

Coordination with other Models

The Cradle to Cradle model can be viewed as a framework that considers systems as a whole or holistically. It can be applied to many aspects of human society, and is related to Life cycle assessment. See for instance the LCA based model of the Eco-costs, which has been designed to cope with analyses of recycle systems. The Cradle to Cradle model in some implementations is closely linked with the Car-free movement, such as in the case of large-scale building projects or the construction or redevelopment of urban environments. It is closely linked with passive solar design in the building industry and with permaculture in agriculture within or near urban environments. An earthship is a perfect example where different re-use models are used, cradle to cradle and permaculture.

In 2005, IE Business School in Madrid launched the Center for Eco-Intelligent Innovation in collaboration with William McDonough to study the implementation of Cradle to Cradle design approaches in pioneering businesses. The academic research of companies lead to the elaboration of the Biosphere Rules, a set of five principles derived from nature that guide the implementation of circular models in production.

Constraints

A major constraint in the optimal recycling of materials is that at civic amenity sites, products are not disassembled by hand and have each individual part sorted into a bin, but instead have the entire product sorted into a certain bin.

This makes the extraction of rare earth elements and other materials uneconomical (at recycling sites, products typically get crushed after which the materials are extracted by means of magnets, chemicals, special sorting methods,) and thus optimal recycling of, for example metals is impossible (an optimal recycling method for metals would require to sort all similar alloys together rather than mixing plain iron with alloys).

Obviously, disassembling products is not feasible at currently designed civic amenity sites, and a better method would be to send back the broken products to the manufacturer, so that the manufacturer can disassemble the product. These disassembled product can then be used for making new products or at least to have the components sent separately to recycling sites (for proper recycling, by the exact type of material). At present though, few laws are put in place in any country to oblige manufacturers to take back their products for disassembly, nor are there even such obligations for manufacturers of cradle-to-cradle products. One process where this is happening is in the EU with the Waste Electrical and Electronic Equipment Directive.

Criticism and Response

Criticism has been advanced on the fact that McDonough and Braungart previously kept C2C consultancy and certification in their inner circle. Critics argued that this lack of competition prevented the model from fulfilling its potential. Many critics pleaded for a public-private partnership overseeing the C2C concept, thus enabling competition and growth of practical applications and services.

McDonough and Braungart responded to this criticism by giving control of the certification protocol to a non-profit, independent Institute called the Cradle to Cradle Products Innovation Institute. McDonough said the new institute "will enable our protocol to become a public certification program and global standard." The new Institute announced the creation of a Certification Standards Board in June 2012. The new board, under the auspices of the Institute, will oversee the certification moving forward.

Experts in the field of environment protection have questioned the practicability of the concept. Friedrich Schmidt-Bleek, head of the German Wuppertal Institute called his assertion, that the "old" environmental movement had hindered innovation with its pessimist approach "pseudo-psychological humbug".

"I can feel very nice on Michael's seat covers in the airplane. Nevertheless I am still waiting for a detailed proposal for a design of the other 99.99 percent of the Airbus 380 after his principles."

In 2009 Schmidt-Bleek stated that it is out of the question that the concept can be realized on a bigger scale.

Some claim that C2C certification may not be entirely sufficient in all eco-design approaches. Quantitative methodologies (LCAs) and more adapted tools (regarding the product type which is considered) could be used in tandem. The C2C concept ignores the use phase of a product. According to the Variants of Life Cycle Assessment the entire life cycle of a product or service has to be evaluated, not only the material itself. For many goods e.g. in transport, the use phase has the most influence on the environmental footprint. E.g. the more lightweight a car or a plane the less fuel it consumes and consequently the less impact it has. Braungart fully ignores the use phase.

It is safe to say that every production step or resource-transformation step needs a certain amount of energy.

The C2C concept foresees an own certification of its analysis and therefore is in contradiction to international ISO standards 14040 and 14044 for Life Cycle Assessment whereas an independent and critical review is needed in order to obtain comparative and resilient results. Independent external review.

Ecodesign

Ecodesign is an approach to designing products with special consideration for the environmental impacts of the product during its whole lifecycle. In a life cycle assessment, the life cycle of a product is usually divided into procurement, manufacture, use, and disposal.

Ecodesign is a growing responsibility and understanding of our ecological footprint on the planet. Green awareness, overpopulation, industrialization and an increased environmental population have led to the questioning of consumer values. It is imperative to search for new building solutions that are environmentally friendly and lead to a reduction in the consumption of materials and energy.

Overview

As the whole product's life cycle should be regarded in an integrated perspective, representatives from advance development, design, production, marketing, purchasing, and project management should work together on the Ecodesign of a further developed or new product. Together, they have the best chance to predict the holistic effects of changes of the product and their environmental impact.

An eco-design product has a cradle-to-cradle life cycle ensuring zero waste is created in the whole process. By mimicking life cycles in nature, eco-design is a fundamental concept in achieving a truly circular economy.

Stainless steel table with FSC Teca wood - Brazil ecodesign

Environmental aspects which ought to be analysed for every stage of the life cycle are:

- Consumption of resources (energy, materials, water or land area)
- Emissions to air, water, and the ground (our Earth) as being relevant for the environment and human health
- Miscellaneous (e.g. noise and vibration)

Waste (hazardous waste and other waste defined in environmental legislation) is only an intermediate step and the final emissions to the environment (e.g. methane and leaching from landfills) are inventoried. All consumables, materials and parts used in the life cycle phases are accounted for, and all indirect environmental aspects linked to their production.

The environmental aspects of the phases of the life cycle are evaluated according to their environmental impact on the basis of a number of parameters, such as extent of environmental impact, potential for improvement, or potential of change.

According to this ranking the recommended changes are carried out and reviewed after a certain time.

Environmental Effect Analysis

An electric wire reel reused as a center table in a Rio de Janeiro decoration fair. The reuse of materials is a sustainable practice that is rapidly growing among designers in Brazil.

One instrument to identify the factors that are important for the reduction of the environmental impact during all lifecycle stages is the environmental effect analysis (EEA).

For an EEA the following are taken into account:

- Customers' wishes
- Legal requirements, market requirements (competitors)
- Data concerning the product and the manufacturing process

Applications in Design

Ecodesign concepts currently have a great influence on many aspects of design; the impact of global warming and an increase in CO_2 emissions have led companies to consider a more environmentally conscious approach to their design thinking and process. In building design and construction, designers are taking on the concept of Ecodesign throughout the design process, from the choice of materials to the type of energy that is being consumed and the disposal of waste.

With respect to these concepts, online platforms dealing in only Ecodesign products are emerging, with the additional sustainable purpose of eliminating all unnecessary distribution steps between the designer and the final customer.

EcoMaterials, such as the use of local raw materials, are less costly and reduce the environmental costs of shipping, fuel consumption, and CO_2 emissions generated from transportation. Certified green building materials, such as wood from sustainably managed forest plantations, with accreditations from companies such as the Forest Stewardship Council (FSC), or the Pan-European Forest Certification Council (PEFCC), can be used.

Several other types of components and materials can be used in sustainable buildings. Recyclable and recycled materials are commonly used in construction, but it is important that they don't generate any waste during manufacture or after their life cycle ends. Reclaimed materials such as timber at a construction site or junkyard can be given a second life by reusing them as support beams in a new building or as furniture. Stones from an excavation can be used in a retaining wall. The reuse of these items means that less energy is consumed in making new products and a new natural aesthetic quality is achieved.

Water recycling systems such as rainwater tanks that harvest water for multiple purposes. Reusing grey water generated by households are a useful way of not wasting drinking water.

Off-grid homes only use clean electric power. They are completely separated and disconnected from the conventional electricity grid and receive their power supply by harnessing active or passive energy systems.

Active System

These systems use the principle of harnessing the power generated from renewable and inexhaustible sources of energy, for example; solar, wind, thermal, biomass, and geothermal energy.

Solar power is a widely known and used renewable energy source. An increase in technology has allowed solar power to be used in a wide variety of applications. Two types of solar panels generate heat into electricity. Thermal solar panels reduce or eliminate the consumption of gas and diesel, and reduce CO_2 emissions. Photovoltaic panels convert solar radiation into an electric current which can power any appliance. However, this is a more complex technology and is generally more expensive to manufacture than thermal panels.

Biomass is the energy source created from organic materials generated through a forced or spontaneous biological process.

Geothermal energy is obtained by harnessing heat from the ground. This type of energy can be used to heat and cool homes. It eliminates dependence on external energy and generates minimum waste. It is also hidden from view as it is placed underground, making it more aesthetically pleasing and easier to incorporate in a design.

Wind turbines are a useful application for areas without immediate conventional power sources, e.g., rural areas with schools and hospitals that need more power. Wind turbines can provide up to 30% of the energy consumed by a household but they are subject to regulations and technical specifications, such as the maximum distance at which the facility is located from the place of consumption and the power required and permitted for each property.

Passive Systems

Buildings that integrate passive energy systems (bioclimatic buildings) are heated using non-mechanical methods, thereby optimizing natural resources. The use of optimal daylight plays an integral role in passive energy systems. This involves the positioning and location of a building to allow and make use of sunlight throughout the whole year. By using the sun's rays, thermal mass is stored into the building materials such as concrete and can generate enough heat for a room.

A green roof is a roof partially or completely covered with plants or other vegetation. This creates insulation that helps regulate the building's temperature. It also retains water, providing a water recycling system. It also provides soundproofing.

In Art and Decorating

Recycling has been used in art since the early part of the 20th century, when cubist artist Pablo Picasso (1881–1973) and Georges Braque (1882–1963) created collages from newsprints, packaging and other found materials. The "Outside Art" movement is recognized as a genuine expressive art form, and is celebrated because of the materials used and not in spite of them. The same principle can be used inside the home, where found objects are now displayed with pride and collecting certain objects and materials to furnish a home is now admired rather than looked down upon.

There is a huge demand in Western countries to decorate homes in a "green" style. A lot of effort is placed into recycled product design and the creation of a natural look. This ideal is also a part of developing countries, although their use of recycled and natural products is often based in necessity and wanting to get maximum use out of materials.

Eco-industrial Park

An eco-industrial park (EIP) is an industrial park in which businesses cooperate with each other and with the local community in an attempt to reduce waste and pollution, efficiently share resources (such as information, materials, water, energy, infrastructure, and natural resources), and help achieve sustainable development, with the intention of increasing economic gains and improving environmental quality. An EIP may also be planned, designed, and built in such a way that it makes it easier for businesses to co-operate, and that results in a more financially sound, environmentally friendly project for the developer.

The Eco-industrial Park Handbook states that "An Eco-Industrial Park is a community of manufacturing and service businesses located together on a common property. Members seek enhanced environmental, economic, and social performance through collaboration in managing environmental and resource issues."

Based on the concepts of industrial ecology, collaborative strategies not only include by-product synergy ("waste-to-feed" exchanges), but can also take the form of wastewater cascading, shared logistics and shipping & receiving facilities, shared parking, green technology purchasing blocks, multi-partner green building retrofit, district energy systems, and local education & resource centres. This is an application of a systems approach, in which designs and processes/activities are integrated to address multiple objectives.

EIPs can be developed as greenfield land projects, where the eco-industrial intent is present throughout the planning, design and site construction phases, or developed through retrofits and new strategies in existing industrial developments.

Examples

"Industrial symbiosis" is a related but more limited concept in which companies in a region collaborate to utilize each other's by-products and otherwise share resources. In Kalundborg, Denmark a symbiosis network links a 1500MW coal-fired power plant with the community and other companies. Surplus heat from this power plant is used to heat 3500 local homes in addition to a nearby fish farm, whose sludge is then sold as a fertilizer. Steam from the power plant is sold to Novo Nordisk, a pharmaceutical and enzyme manufacturer, in addition to a Statoil plant. This reuse of heat reduces the amount thermal pollution discharged to a nearby fjord. Additionally, a by-product from the power plant's sulfur dioxide scrubber contains gypsum, which is sold to a wallboard manufacturer. Almost all of the manufacturer's gypsum needs are met this way, which reduces the amount of open-pit mining needed. Furthermore, fly ash and clinker from the power plant is utilized for road building and cement production.

The industrial symbiosis at Kalundborg was not created as a top-down initiative, but instead evolved gradually. As environmental regulations became stricter, firms were motivated reduce the cost of compliance, and turn their by-products into economic products.

In Canada, eco-industrial parks exist across the country and have enjoyed some success. The best known example is Burnside Park, in Halifax, Nova Scotia. With support from Dalhousie University's Eco-Efficiency Centre, the more than 1,500 businesses have been improving their environmental performance and developing profitable partnerships. Subsequently, two greenfield industrial developments have been started in Alberta: TaigaNova Eco-Industrial Park is in the heart of the Athabasca oil sands, while Innovista Eco-Industrial Park is a gateway to the Rocky Mountains ~300km west of Edmonton.

UNIDO Viet Nam (United Nations Industrial Development Organization) has compiled a list in 2015 of Eco-Industrial Parks (EIP) in the ASEAN Economic Community in a report titled "Economic Zones in the ASEAN" written by Arnault Morisson.

Other Usage

EIPs also refer to industrial parks where a "green" approach has been taken towards the infrastructure and development of the site. This can include green infrastructure related to Renewable Energy Systems; stormwater, groundwater and wastewater management; road surfaces; and transportation demand management. Green building practices can also be encouraged or mandated

EIPs are often used as a stimulus for economic diversification in the community or region where they are located. Anchor tenants, such as bio-based product manufacturers or waste-to-energy facilities, etc., can attract complementary businesses as suppliers, scavengers/recyclers, service providers, downstream users and other businesses that could benefit from eco-industrial strategies.

Suggested Usage

It is suggested that EIPs be used as a means of growing the renewable energy sector. In the case of a Solar Photovoltaic (PV) Manufacturing plant, an EIP can increase the manufacturing efficiency to make it more economical, while reducing the environmental impact of producing the solar cells. In essence, this assists the growth of the renewable energy industry and the environmental benefits that come with replacing fossil-fuels.

Integrated Chain Management

Integrated Chain Management (ICM), also known as Integral Chain Management, is an approach for the reduction of environmental impact of product chains. Such a product chain exists out of an extraction phase, a production phase, a use phase and a waste phase. The ultimate goal of ICM is a reduction of environmental load over the whole chain. Integrated Chain Management is one of the approaches that can be used to come to sustainable development. Other approaches in this line are the Ecological Footprint and the DTO approach.

Within the ICM approach all phases within the chain must be considered. Therefore it can be seen as a "cradle to grave" approach. Several inputs and outputs can be taken into account when applying the ICM approach. Such as: Energy flows, mass flows, materials, waste flows and emissions. Within ICM material cycles should be closed where possible and the remainder flows of emissions and waste should be brought within acceptable boundaries. Also the use of resources should be kept to a minimum.

Integrated chain management should not be mixed up with Supply Chain Management or Integrated Supply Chain Management. These concepts do not have the reduction of environmental load as their main goal.

An important aspect of ICM is that shifting to other phases in the product chain is avoided. For instance, a producer of chairs can choose to leave off an environment unfriendly material in a new product. The producer can even see this as an extra selling point for the customer, but as a consequence the supplier of raw materials has to use much more energy to produce a material with the same qualities. The result of this is that there may no longer be a net environmental reduction across the whole chain. Within the integrated chain management approach this is avoided.

The chain can be managed by developing new policies and economical or political incentives. Therefore one must have insight into the inputs and outputs of the production chain. Before these policies can be developed one must engage in several actions.

- Analyse the processes into a preferred level of detail
- Determine the boundaries of the chain. Should links outside the companies be involved as well?
- Determine whether there should be a focus on just one or on several environmental problems

- Determine on which material flows or energy flows there should be a focus.

Effective supply chain management can impact virtually all business and production processes

Example

An example of applying the ICM approach would be to develop policies in a particular product area. The responsibility of problems caused by the waste stage can be assigned to the producers of these products. This leads to improved product design and new insight in how to put these products in the market. For instance the product can be sold with a disposal contribution. On the price tag of a radio nowadays can be printed: "this radio costs 25 $ not including the 3 $ disposal contribution" The effects can be seen within the whole chain. The producer will try to choose non-polluting materials, as they increase the costs of the waste-stage. The producer of raw materials will try to improve its production process in order to meet the increased demand for 'clean' primary products. And the consumer will be aware that some products give more pressure on the environment than others when its economical lifespan has run out.

Environmental Management System

Environmental management system (EMS) refers to the management of an organization's environmental programs in a comprehensive, systematic, planned and documented manner. It includes the organizational structure, planning and resources for developing, implementing and maintaining policy for environmental protection.

More formally, EMS is "a system and database which integrates procedures and processes for training of personnel, monitoring, summarizing, and reporting of specialized environmental performance information to internal and external stakeholders of a firm."

The most widely used standard on which an EMS is based is International Organization for Standardization (ISO) 14001. Alternatives include the EMAS.

An environmental management information system (EMIS) is an information technology solution for tracking environmental data for a company as part of their overall environmental management system.

Goals

The goals of EMS are to increase compliance and reduce waste:

- Compliance is the act of reaching and maintaining minimal legal standards. By not being compliant, companies may face fines, government intervention or may not be able to operate.

- Waste reduction goes beyond compliance to reduce environmental impact. The EMS helps to develop, implement, manage, coordinate and monitor environmental policies. Waste reduction begins at the design phase through pollution prevention and waste minimization. At the end of the life cycle, waste is reduced by recycling.

To meet these goals, the selection of environmental management systems is typically subject to a certain set of criteria: a proven capability to handle high frequency data, high performance indicators, transparent handling and processing of data, powerful calculation engine, customised factor handling, multiple integration capabilities, automation of workflows and QA processes and in-depth, flexible reporting.

Features

An environmental management system (EMS):

- Serves as a tool, or process, to improve environmental performance and information mainly "design, pollution control and waste minimization, training, reporting to top management, and the setting of goals"

- Provides a systematic way of managing an organization's environmental affairs

- Is the aspect of the organization's overall management structure that addresses immediate and long-term impacts of its products, services and processes on the environment. EMS assists with planning, controlling and monitoring policies in an organization.

- Gives order and consistency for organizations to address environmental concerns through the allocation of resources, assignment of responsibility and ongoing evaluation of practices, procedures and processes

- Creates environmental buy-in from management and employees and assigns accountability and responsibility.

- Sets framework for training to achieve objectives and desired performance.

- Helps understand legislative requirements to better determine a product or service's impact, significance, priorities and objectives.

- Focuses on continual improvement of the system and a way to implement policies and objectives to meet a desired result. This also helps with reviewing and auditing the EMS to find future opportunities.

- Encourages contractors and suppliers to establish their own EMS.
- Facilitates e-reporting to federal, state and provincial government environmental agencies through direct upload.

EMS Model

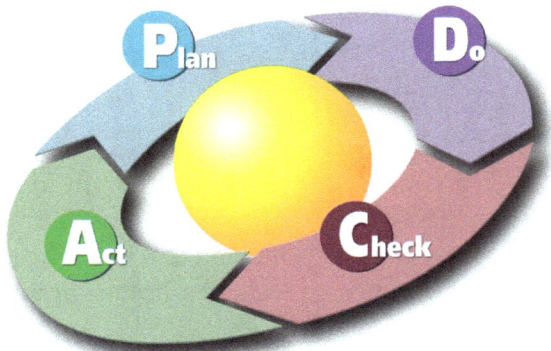

The PDCA cycle

An EMS follows a Plan-Do-Check-Act, or PDCA, Cycle. The diagram shows the process of first developing an environmental policy, planning the EMS, and then implementing it. The process also includes checking the system and acting on it. The model is continuous because an EMS is a process of continual improvement in which an organization is constantly reviewing and revising the system.

This is a model that can be used by a wide range of organizations — from manufacturing facilities to service industries to government agencies.

Other Meanings

An EMS can also be classified as

- a system which monitors, tracks and reports emissions information, particularly with respect to the oil and gas industry. EMSs are becoming web-based in response to the EPA's mandated greenhouse gas (GHG) reporting rule, which allows for reporting GHG emissions information via the internet.
- a centrally controlled and often automated network of devices (now frequently wireless using z-wave and zigbee technologies) used to control the internal environment of a building. Such a system namely acts as an interface between end user and energy (gas/electricity) consumption.

Examples of Environmental Management Systems

- Emisoft's environmental management, reporting and compliance platform

- Medgate environmental management software
- EsDat environmental data management system
- Enviance regulatory compliance system
- ERA Environmental's environmental management system

Ecological Modernization

Ecological modernization is an optimistic school of thought in the social sciences that argues that the economy benefits from moves towards environmentalism. It has gained increasing attention among scholars and policymakers in the last several decades internationally. It is an analytical approach as well as a policy strategy and environmental discourse (Hajer, 1995).

Key Elements

Ecological modernization emerged in the early 1980s within a group of scholars at Free University and the Social Science Research Centre in Berlin, among them Joseph Huber, Martin Jänicke (de) and Udo E. Simonis (de). Various authors pursued similar ideas at the time, e.g. Arthur H. Rosenfeld, Amory Lovins, Donald Huisingh, René Kemp, or Ernst Ulrich von Weizsäcker. Further substantial contributions were made by Arthur P.J. Mol, Gert Spaargaren and David A Sonnenfeld (Mol and Sonnenfeld, 2000; Mol, 2001).

One basic assumption of ecological modernization relates to environmental readaptation of economic growth and industrial development. On the basis of enlightened self-interest, economy and ecology can be favourably combined: Environmental productivity, i.e. productive use of natural resources and environmental media (air, water, soil, ecosystems), can be a source of future growth and development in the same way as labour productivity and capital productivity. This includes increases in energy and resource efficiency as well as product and process innovations such as environmental management and sustainable supply chain management, clean technologies, benign substitution of hazardous substances, and product design for environment. Radical innovations in these fields can not only reduce quantities of resource turnover and emissions, but also change the quality or structure of the industrial metabolism. In the co-evolution of humans and nature, and in order to upgrade the environment's carrying capacity, ecological modernization gives humans an active role to play, which may entail conflicts with nature conservation.

There are different understandings of the scope of ecological modernization - whether it is just about techno-industrial progress and related aspects of policy and economy, and to what extent it also includes cultural aspects (ecological modernization of mind, value orientiations, attitudes, behaviour and lifestyles). Similarly, there is some pluralism as to whether ecological modernization would need to rely mainly on government, or markets and entrepreneurship, or civil society, or some sort of multi-level governance combining the three. Some scholars explicitly refer to general modernization theory as well as non-Marxist world-system theory, others don't.

Ultimately, however, there is a common understanding that ecological modernization will have to result in innovative structural change. So research is now still more focused on environmental innovations, or eco-innovations, and the interplay of various societal factors (scientific, economic, institutional, legal, political, cultural) which foster or hamper such innovations (Klemmer et al., 1999; Huber, 2004; Weber and Hemmelskamp, 2005; Olsthoorn and Wieczorek, 2006).

Ecological modernization shares a number of features with neighbouring, overlapping approaches. Among the most important are

- the concept of sustainable development
- the approach of industrial metabolism (Ayres and Simonis, 1994)
- the concept of industrial ecology (Socolow, 1994).

Additional Elements

A special topic of ecological modernization research during recent years was *sustainable household*, i.e. environment-oriented reshaping of lifestyles, consumption patterns, and demand-pull control of supply chains (Vergragt, 2000; OECD 2002). Some scholars of ecological modernization share an interest in industrial symbiosis, i.e. inter-site recycling that helps to reduce the consumption of resources via increasing efficiency (i.e. pollution prevention, waste reduction), typically by taking externalities from one economic production process and using them as raw material inputs for another (Christoff, 1996). Ecological modernization also relies on product life-cycle assessment and the analysis of materials and energy flows. In this context, ecological modernization promotes 'cradle to cradle' manufacturing (Braungart and McDonough, 2002), contrasted against the usual 'cradle to grave' forms of manufacturing - where waste is not re-integrated back into the production process. Another special interest in the ecological modernization literature has been the role of social movements and the emergence of civil society as a key agent of change (Fisher and Freudenburg, 2001).

As a strategy of change, some forms of ecological modernization may be favored by business interests because they seemingly meet the triple bottom line of economics, society, and environment, which, it is held, underpin sustainability, yet do not challenge free market principles. This contrasts with many environmental movement perspectives, which regard free trade and its notion of business self-regulation as part of the problem, or even an origin of environmental degradation. Under ecological modernization, the state is seen in a variety of roles and capacities: as the enabler for markets that help produce the technological advances via competition; as the regulatory medium through which corporations are forced to 'take back' their various wastes and re-integrate them in some manner into the production of new goods and services (e.g. the way that car corporations in Germany are required to accept back cars they manufactured once those vehicles have reached the end of their product lifespan); and in some cases as an institution that is incapable of addressing critical local, national, and global environmental problems. In the latter case, ecological modernization shares with Ulrich Beck (1999, 37-40) and others notions of the necessity of emergence of new forms of environmental governance, sometimes referred to as subpolitics or political modernization, where the environmental movement, community groups, businesses, and other stakeholders increasingly take on direct and leadership roles in stimulating environmental transformation. Political modernization of this sort requires certain supporting norms and institutions such as a free, independent, or at least critical press, basic human rights of expression, organization, and assembly, etc. New media such as the Internet greatly facilitate this.

Criticisms

Critics argue that ecological modernization will fail to protect the environment and does nothing to alter the impulses within the capitalist economic mode of production that inevitably lead to environmental degradation (Foster, 2002). As such, it is just a form of 'green-washing'. Critics question whether technological advances alone can achieve resource conservation and better environmental protection, particularly if left to business self-regulation practices (York and Rosa, 2003). For instance, many technological improvements are currently feasible but not widely utilized. The most environmentally friendly product or manufacturing process (which is often also the most economically efficient) is not always the one automatically chosen by self-regulating corporations (e.g. hydrogen or biofuel vs. peak oil). In addition, some critics have argued that ecological modernization does not redress gross injustices that are produced within the capitalist system, such as environmental racism - where people of color and low income earners bear a disproportionate burden of environmental harm such as pollution, and lack access to environmental benefits such as parks, and social justice issues such as eliminating unemployment (Bullard, 1993; Gleeson and Low, 1999; Harvey, 1996) - environmental racism is also referred to as issues of the asymmetric distribution of environmental resources and services (Everett & Neu, 2000). Moreover, the theory seems to have limited global efficacy, applying primarily to its countries of origin - Germany and the Netherlands, and having little to say about the developing world (Fisher and Freudenburg, 2001). Perhaps the harshest criticism though, is that ecological modernization is predicated upon the notion of 'sustainable growth', and in reality this is not possible because growth entails the consumption of natural and human capital at great costs to ecosystems and societies.

Ecological modernization, its effectiveness and applicability, strengths and limitations, remains a dynamic and contentious area of environmental social science research and policy discourse in the early 21st century.

Regenerative Design

Regenerative design is a process-oriented systems theory based approach to design. The term "regenerative" describes processes that restore, renew or revitalize their own sources of energy and materials, creating sustainable systems that integrate the needs of society with the integrity of nature. The basis is derived from systems ecology with a closed loop input–output model or a model in which the output is greater than or equal to the input with all outputs viable and all inputs accounted for. Regenerative design is the biomimicry of ecosystems that provide for all human systems to function as a closed viable ecological economics system for all industry. It parallels ecosystems in that organic (biotic) and synthetic (abiotic) material is not just metabolized but metamorphosed into new viable materials. Ecosystems and regeneratively designed systems are holistic frameworks that seek to create systems that are absolutely waste free. The model is meant to be applied to many different aspects of human habitation such as urban environments, buildings, economics, industry and social systems. Simply put, it is the design of ecosystems and human behavior, or culture that function as human habitats.

Whereas the highest aim of sustainable development is to satisfy fundamental human needs today without compromising the possibility of future generations to satisfy theirs, the end-goal of regenerative design is to redevelop systems with absolute effectiveness, that allows for the co-evolution of the human species along with other thriving species.

History

During the late 1970s, John T. Lyle (1934–1998), a landscape architecture professor, challenged graduate students to envision a community in which daily activities were based on the value of living within the limits of available renewable resources without environmental degradation. Over the next few decades an eclectic group of students, professors and experts from around the world and crossing many disciplines developed designs for an institute to be built at Cal Poly Pomona. In 1992 the Lyle Center for Regenerative Studies was built over two years and opened in 1994. In that same year Lyle's book *Regenerative Design for Sustainable Development* was published by Wiley. In 1995 Lyle worked with William McDonough at Oberlin College for the Adam Joseph Lewis Center for Environmental Studies completed in 2000. In 2002 McDonough's book, the more popular and successful, *Cradle to Cradle: Remaking the Way We Make Things* was published reiterating the concepts developed by Lyle.

Lyle saw the connection between concepts developed by Bob Rodale of the Rodale Institute for regenerative agriculture and the opportunity to develop regenerative systems for all other aspects of the world. While regenerative agriculture focused solely on agriculture, Lyle expanded its concepts and use to all systems. With regenerative agriculture, the concepts are very straight forward and simple but Lyle understood that when developing for other types of systems, more complicated ideas such as entropy and emergy must be taken into consideration.

Swiss architect Walter R. Stahel developed approaches entirely similar to Lyle's also in the late 1970s but instead coined the term cradle-to-cradle design made popular by McDonough and Michael Braungart

Regenerative Versus Sustainable

Regenerative and sustainable are essentially the same thing except for one key point: in a sustainable system, lost ecological systems are not returned to existence. In a regenerative system, those lost systems can ultimately begin "regenerating" back into existence. Put more simply, regenerative systems create a "better" world than we (humans) found it, now and into the future.

There is also a linguistic problem with the word "sustainable", which in the strict sense is meant to mean "self-sustaining". Because the word root "sustain" means "last" or "endure," the general public and even many non-experts in the industry define the word only as "able to last" or "the capacity to endure." In popular usage by designers and product manufacturers, "sustainable" has become a relative term referring to any material, process or product (including a building) which is less toxic or environmental harmful than those conventionally used. A product that contains 75% recycled material is often considered "sustainable", but is in fact merely MORE sustainable than a comparable product that contains no recycled material. A truly sustainable material would be one made of 100% recycled material that can, in turn, be completely recycled into a comparable new material or product. This is rarely the case.

"Regenerative" also suffers from a slightly different linguistic problem, the term is still competing with the biological community in terms of its use for the re-growth of limbs etc.. However once the word itself gains wide usage, its meaning becomes more general, much like in the case of the term "sustainable". The base meaning of "re-generative" means the "capacity to bring into existence again." So if an item or system is regenerative the item or system has the capacity to bring itself into existence again. Using the example above, a truly regenerative product would not only be 100% recycled and recyclable, but it would also improve the environmental conditions at the factory where it was made, the business where it was used and so on throughout its life-cycle (creating habitat, filtering water, catalyzing nitrogen-fixation processes in the soil, etc.).

Preservation Versus Conservation

Regenerists place more importance on conservation than on preservation. It is recognized in regenerative design that humans are a part of natural ecosystems. To exclude people is to create dense areas that destroy pockets of existing ecosystems while preserving pockets of ecosystems without allowing them to change naturally over time. By incorporating people into ecosystems all inputs are pulled from local areas and all outputs are accounted for creating a waste-less system. When human systems cease to create waste, what would once have been considered waste becomes a resource for the input in which the output comes from.

Food Systems

Regenerists call for the creation of demand on agricultural systems to produce regenerative foods. This is often compared to the creation of the demand for organic food. Organic foods have a relation to regenerative foods in that regenerative food is all organic, but not all organic food is regenerative. Organic food is not regenerative if the byproduct of the food crop is not a resource for the next seasons crops and if other inputs for the crop did not come from other resources within the farm which it is grown in.

Size of Regenerative Systems

The size of the regenerative system effects its regenerativity. The smaller a system is designed the more likely it is to be stable and regenerative. Multiple small regenerative systems that are put together to create larger regenerative systems help to create supplies for multiple human-inclusive-ecological systems.

Quantifying Regenerativity

Due to evolution and the continuing and largely unpredictable changes that occur over the lifetime of Earth, it is impossible to create a 100% regenerative system. One can only reach 99.999% efficiency, the ultimate goal. However, with the energy material interchange, it is possible to create enough energy to potentially create the equivalent amount of material used to create the system in the first instance.

A completed object (an object with emergy, or embodied energy) can however create more energy than was used to create it. I.e. a solar panel outputting more energy than its given embodied energy. However the system used to make up the solar panel: the inputs such as the materials for the object (silicon) and the solar radiation can only be regenerated if enough energy is produced to generate the materials used to make up the solar panel. However, today's solar panels are not yet that energy efficient.

References

- McAloone, T. C. & Bey, N. (2009), Environmental improvement through product development - a guide, Danish EPA, Copenhagen Denmark, ISBN 978-87-7052-950-1
- The Journal of Design History: Environmental conscious design and inverse manufacturing,2005. Eco Design 2005, 4th International Symposium
- Hajer, M.A., 1995, The Politics of Environmental Discourse: Ecological Modernization and the Policy Process, Oxford, UK, Oxford University Press, ISBN 0-19-827969-8
- Christoff, Peter (1996). "Ecological modernisation, ecological modernities". Environmental Politics. Informa UK Limited. 5 (3): 476–500. ISSN 0964-4016. doi:10.1080/09644019608414283
- Mol, A.P.J., 2001, Globalization and Environmental Reform: The Ecological Modernization of the Global Economy, Cambridge, Ma., MIT Press, ISBN 0-262-13395-4
- The Design Journal: Vol 13, Number 1, March 2010 - Design is the problem: The future of Design must be sustainable, N. Shedroff
- Redclift, M. R., and Woodgate, G., (eds.) 2005, New Developments in Environmental Sociology, Cheltenham, Edward Elgar, ISBN 1-84376-115-7
- York, Richard; Rosa, Eugene A. (2003-09-01). "Key Challenges to Ecological Modernization Theory: Institutional Efficacy, Case Study Evidence, Units of Analysis, and the Pace of Eco-Efficiency". Organization & Environment. SAGE Publications. 16 (3): 273–288. doi:10.1177/1086026603256299

Methods and Techniques of Industrial Ecology

The methods and techniques of industrial ecology are SWOT analysis, ecological footprint, energy accounting, zero waste, eco-costs value ratio and rebound effect. SWOT analysis is carried out for companies and products. It can be categorized into internal factors and external factors. This chapter discusses the methods of industrial ecology in a critical manner providing key analysis to the subject matter.

SWOT Analysis

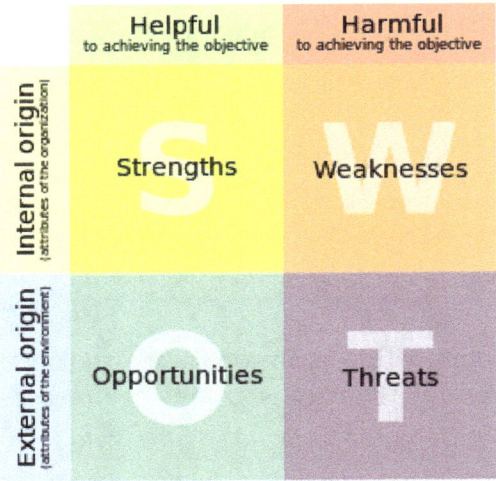

A SWOT analysis, with its four elements in a 2×2 matrix.

SWOT analysis (alternatively SWOT matrix) is an acronym for *strengths, weaknesses, opportunities,* and *threats* and is a structured planning method that evaluates those four elements of an organization, project or business venture. A SWOT analysis can be carried out for a company, product, place, industry, or person. It involves specifying the objective of the business venture or project and identifying the internal and external factors that are favorable and unfavorable to achieve that objective. Some authors credit SWOT to Albert Humphrey, who led a convention at the Stanford Research Institute (now SRI International) in the 1960s and 1970s using data from Fortune 500 companies. However, Humphrey himself did not claim the creation of SWOT, and the origins remain obscure. The degree to which the internal environment of the firm matches with the external environment is expressed by the concept of strategic fit.

- Strengths: characteristics of the business or project that give it an advantage over others
- Weaknesses: characteristics of the business that place the business or project at a disadvantage relative to others
- Opportunities: elements in the environment that the business or project could exploit to its advantage
- Threats: elements in the environment that could cause trouble for the business or project

Identification of SWOTs is important because they can inform later steps in planning to achieve the objective. First, decision-makers should consider whether the objective is attainable, given the SWOTs. If the objective is *not* attainable, they must select a different objective and repeat the process.

Users of SWOT analysis must ask and answer questions that generate meaningful information for each category (strengths, weaknesses, opportunities, and threats) to make the analysis useful and find their competitive advantage.

Internal and External Factors

"So it is said that if you know your enemies and know yourself, you can win a hundred battles without a single loss. If you only know yourself, but not your opponent, you may win or may lose. If you know neither yourself nor your enemy, you will always endanger yourself."

The Art of War by Sun Tzu

SWOT analysis aims to identify the key internal and external factors seen as important to achieving an objective. SWOT analysis groups key pieces of information into two main categories:

- Internal factors – the *strengths* and *weaknesses* internal to the organization
- External factors – the *opportunities* and *threats* presented by the environment external to the organization

Analysis may view the internal factors as strengths or as weaknesses depending upon their effect on the organization's objectives. What may represent strengths with respect to one objective may be weaknesses (distractions, competition) for another objective. The factors may include all of the 4Ps as well as personnel, finance, manufacturing capabilities, and so on.

The external factors may include macroeconomic matters, technological change, legislation, and sociocultural changes, as well as changes in the marketplace or in competitive position. The results are often presented in the form of a matrix.

SWOT analysis is just one method of categorization and has its own weaknesses. For example, it may tend to persuade its users to compile lists rather than to think about actual important factors in achieving objectives. It also presents the resulting lists uncritically and without clear prioritization so that, for example, weak opportunities may appear to balance strong threats.

It is prudent not to eliminate any candidate SWOT entry too quickly. The importance of individual SWOTs will be revealed by the value of the strategies they generate. A SWOT item that produces valuable strategies is important. A SWOT item that generates no strategies is not important.

Use

The usefulness of SWOT analysis is not limited to profit-seeking organizations. SWOT analysis may be used in any decision-making situation when a desired end-state (objective) is defined. Examples include non-profit organizations, governmental units, and individuals. SWOT analysis may also be used in pre-crisis planning and preventive crisis management. SWOT analysis may also be used in creating a recommendation during a viability study/survey.

Strategy Building

SWOT analysis can be used effectively to build organizational or personal strategy. Steps necessary to execute strategy-oriented analysis involve identification of internal and external factors (using popular the 2x2 matrix), selection and evaluation of the most important factors, and identification of relations existing between internal and external features.

For instance, strong relations between strengths and opportunities can suggest good conditions in the company and allow using an *aggressive* strategy. On the other hand, strong interactions between weaknesses and threats could be analyzed as a potential warning and advice for using a *defensive* strategy.

Matching and Converting

One way of utilizing SWOT is matching and converting. Matching is used to find competitive advantage by matching the strengths to opportunities. Another tactic is to convert weaknesses or threats into strengths or opportunities. An example of a conversion strategy is to find new markets. If the threats or weaknesses cannot be converted, a company should try to minimize or avoid them.

Corporate Planning

As part of the development of strategies and plans to enable the organization to achieve its objectives, that organization will use a systematic/rigorous process known as corporate planning. SWOT alongside PEST/PESTLE can be used as a basis for the analysis of business and environmental factors.

- Set objectives – defining what the organization is going to do
- Environmental scanning
    - Internal appraisals of the organization's SWOT, this needs to include an assessment of the present situation as well as a portfolio of products/services and an analysis of the product/service life cycle
- Analysis of existing strategies, this should determine relevance from the results of an internal/external appraisal. This may include gap analysis of environmental factors
- Strategic Issues defined – key factors in the development of a corporate plan that the organization must address
- Develop new/revised strategies – revised analysis of strategic issues may mean the objectives need to change
- Establish critical success factors – the achievement of objectives and strategy implementation
- Preparation of operational, resource, projects plans for strategy implementation
- Monitoring results – mapping against plans, taking corrective action, which may mean amending objectives/strategies

Marketing

In many competitor analyses, marketers build detailed profiles of each competitor in the market, focusing especially on their relative competitive strengths and weaknesses using SWOT analysis. Marketing managers will examine each competitor's cost structure, sources of profits, resources and competencies, competitive positioning and product differentiation, degree of vertical integration, historical responses to industry developments, and other factors.

Marketing management often finds it necessary to invest in research to collect the data required to perform accurate marketing analysis. Accordingly, management often conducts market research (alternately marketing research) to obtain this information. Marketers employ a variety of techniques to conduct market research, but some of the more common include:

- Qualitative marketing research such as focus groups
- Quantitative marketing research such as statistical surveys
- Experimental techniques such as test markets
- Observational techniques such as ethnographic (on-site) observation
- Marketing managers may also design and oversee various environmental scanning and competitive intelligence processes to help identify trends and inform the company's marketing analysis.

Below is an example SWOT analysis of a market position of a small management consultancy with specialism in HRM.

Strengths	Weaknesses	Opportunities	Threats
Reputation in marketplace	Shortage of consultants at operating level rather than partner level	Well established position with a well-defined market niche	Large consultancies operating at a minor level
Expertise at partner level in HRM consultancy	Unable to deal with multidisciplinary assignments because of size or lack of ability	Identified market for consultancy in areas other than HRM	Other small consultancies looking to invade the marketplace

In Community Organization

		Strengths 1. 2. 3. 4.	Weaknesses 1. 2. 3. 4.
	Opportunities 1. 2. 3. 4.	Opportunity-Strength strategies *Use strengths to take advantage of opportunities* 1. 2.	Opportunity-Weakness strategies *Overcome weaknesses by taking advantage of opportunities* 1. 2.
	Threats 1. 2. 3. 4.	Threat-Strength strategies *Use strengths to avoid threats* 1. 2.	Threat-Weakness Strategies *Minimize weaknesses and avoid threats* 1. 2.

one example of a SWOT Analysis used in community organizing

The SWOT analysis has been utilized in community work as a tool to identify positive and negative factors within organizations, communities, and the broader society that promote or inhibit successful implementation of social services and social change efforts. It is used as a preliminary resource, assessing strengths, weaknesses, opportunities, and threats in a community served by a nonprofit or community organization. This organizing tool is best used in collaboration with community workers and/or community members before developing goals and objectives for a program design or implementing an organizing strategy. The SWOT analysis is a part of the planning for social change process and will not provide a strategic plan if used by itself. After a SWOT analysis is completed, a social change organization can turn the SWOT list into a series of recommendations to consider before developing a strategic plan.

SWOT ANALYSIS

Internal		External	
Strengths	Weaknesses	Opportunities	Threats

A simple SWOT Analysis used in Community Organizing

Strengths and Weaknesses: *These are the internal factors within an organization.*

- Human resources - staff, volunteers, board members, target population
- Physical resources - your location, building, equipment
- Financial - grants, funding agencies, other sources of income
- Activities and processes - programs you run, systems you employ
- Past experiences - building blocks for learning and success, your reputation in the community

Opportunities and Threats: *These are external factors stemming from community or societal forces.*

- Future trends in your field or the culture
- The economy - local, national, or international
- Funding sources - foundations, donors, legislatures
- Demographics - changes in the age, race, gender, culture of those you serve or in your area
- The physical environment (Is your building in a growing part of town? Is the bus company cutting routes?)

- Legislation (Do new federal requirements make your job harderor easier?)
- Local, national, or international events

Although the SWOT analysis was originally designed as an organizational method for business and industries, it has been replicated in various community work as a tool for identifying external and internal support to combat internal and external opposition. The SWOT analysis is necessary to provide direction to the next stages of the change process. It has been utilized by community organizers and community members to further social justice in the context of Social Work practice.

Application in Community Organization

Elements to Consider

Elements to consider in a SWOT analysis include understanding the community that a particular organization is working with. This can be done via public forums, listening campaigns, and informational interviews. Data collection will help inform the community members and workers when developing the SWOT analysis. A needs and assets assessment are tooling that can be used to identify the needs and existing resources of the community. When these assessments are done and data has been collected, an analysis of the community can be made that informs the SWOT analysis.

Steps for Implementation

A SWOT analysis is best developed in a group setting such as a work or community meeting. A facilitator can conduct the meeting by first explaining what a SWOT analysis is as well as identifying the meaning of each term.

One way of facilitating the development of a SWOT analysis includes developing an example SWOT with the larger group then separating each group into smaller teams to present to the larger group after set amount of time. This allows for individuals, who may be silenced in a larger group setting, to contribute. Once the allotted time is up, the facilitator may record all the factors of each group onto a large document such as a poster board, and then the large group, as a collective, can go work through each of the threats and weaknesses to explore options that may be used to combat negative forces with the strengths and opportunities present within the organization and community. A SWOT meeting allows participants to creatively brainstorm, identify obstacles, and possibly strategize solutions/way forward to these limitations.

When to Use SWOT Analysis

The uses of a SWOT analysis by a community organization are as follows: to organize information, provide insight into barriers that may be present while engaging in social change processes, and identify strengths available that can be activated to counteract these barriers.

A SWOT analysis can be used to:

- Explore new solutions to problems
- Identify barriers that will limit goals/objectives
- Decide on direction that will be most effective
- Reveal possibilities and limitations for change
- To revise plans to best navigate systems, communities, and organizations
- As a brainstorming and recording device as a means of communication
- To enhance "credibility of interpretation" to be utilized in presentation to leaders or key supporters.

Benefits

The SWOT analysis in social work practice framework is beneficial because it helps organizations decide whether or not an objective is obtainable and therefore enables organizations to set achievable goals, objectives, and steps to further the social change or community development effort. It enables organizers to take visions and produce practical and efficient outcomes that effect long-lasting change, and it helps organizations gather meaningful information to maximize their potential. Completing a SWOT analysis is a useful process regarding the consideration of key organizational priorities, such as gender and cultural diversity and fundraising objectives.

Limitations

Some findings from Menon et al. (1999) and Hill and Westbrook (1997) have suggested that SWOT may harm performance and that "no-one subsequently used the outputs within the later stages of the strategy".

Other critiques include the misuse of the SWOT analysis as a technique that can be quickly designed without critical thought leading to a misrepresentation of strengths, weaknesses, opportunities, and threats within an organization's internal and external surroundings.

Another limitation includes the development of a SWOT analysis simply to defend previously decided goals and objectives. This misuse leads to limitations on brainstorming possibilities and "real" identification of barriers. This misuse also places the organization's interest above the well-being of the community. Further, a SWOT analysis should be developed as a collaborative with a variety of contributions made by participants including community members. The design of a SWOT analysis by one or two community workers is limiting to the realities of the forces, specifically external factors, and devalues the possible contributions of community members.

Zero Waste

Used products dumped in a landfill

Zero Waste is a philosophy that encourages the redesign of resource life cycles so that all products are reused. No trash is sent to landfills or incinerators. The process recommended is one similar to the way that resources are reused in nature. The definition adopted by the *Zero Waste International Alliance* (ZWIA) is:

Zero Waste is a goal that is ethical, economical, efficient and visionary, to guide people in changing their lifestyles and practices to emulate sustainable natural cycles, where all discarded materials are designed to become resources for others to use.

Zero Waste means designing and managing products and processes to systematically avoid and eliminate the volume and toxicity of waste and materials, conserve and recover all resources, and not burn or bury them.

Implementing Zero Waste will eliminate all discharges to land, water or air that are a threat to planetary, human, animal or plant health

Zero Waste refers to waste management and planning approaches which emphasize waste prevention as opposed to end-of-pipe waste management. It is a whole systems approach that aims for a massive change in the way materials flow through society, resulting in no waste. Zero waste encompasses more than eliminating waste through recycling and reuse, it focuses on restructuring production and distribution systems to reduce waste. Zero waste is more of a goal or ideal rather than a hard target. Zero Waste provides guiding principles for continually working towards eliminating wastes.

Advocates expect that government regulation is needed to influence industrial choices over product and packaging design, manufacturing processes, and material selection.

Advocates say eliminating waste eliminates pollution, and can also reduce costs due to reduced need for raw materials.

Cradle-to-cradle / cradle-to-grave

Cradle-to-grave is a term used to describe a linear model for materials that begins with resource extraction, moves to product manufacturing, and, ends with a 'grave', where the product is disposed of in a landfill. Cradle-to-grave is in direct contrast to cradle-to-cradle. Cradle-to-cradle is a term used in life-cycle analysis to describe a material or product that is recycled into a new product at the end of its life, so that ultimately there is no waste.

Cradle-to-cradle focuses on designing industrial systems so that materials flow in closed loop cycles which mean that waste is minimized, and waste products can be recycled and reused. Cradle-to-cradle simply goes beyond dealing with issues of waste after it has been created, by addressing problems at the source and by re-defining problems by focusing on design. The cradle-to-cradle model is sustainable and considerate of life and future generations.

The cradle-to-cradle framework has evolved steadily from theory to practice. In the industrial sector, it is creating a new notion of materials and material flows. Just as in the natural world, in which one organism's 'waste', cycles through an ecosystem to provide nourishment for other living things, cradle-to-cradle materials circulate in closed-loop cycles, providing nutrients for nature or industry.

An example of a closed loop, cradle-to-cradle product design is DesignTex Fabric. It has designed an upholstery fabric, Climatex Lifecycle, which is a blend of pesticide- and residue-free wool and organically grown ramie, dyed and processed entirely with nontoxic chemicals. All of its product and process inputs were defined and selected for their human and ecological safety within the biological metabolism. This allows the fabric trimmings to be made into felt and used by garden clubs as mulch for growing fruits and vegetables, returning the textile's biological nutrients to the soil.

Benefits

There is a growing global population that is faced with limited resources from the environment. To relieve the pressures placed on the finite resources available it has become more important to prevent waste. To achieve zero waste, waste management has to move from a linear system to being more cyclical so that materials, products and substances are used as efficiently as possible. Materials must be chosen so that it may either return safely to a cycle within the environment or remain viable in the industrial cycle.

Zero waste promotes not only reuse and recycling, but, more importantly, it promotes prevention and product designs that consider the entire product life cycle. Zero waste designs strive for reduced materials use, use of recycled materials, use of more benign materials, longer product lives, reparability, and ease of disassembly at end of life. Zero waste strongly supports sustainability by protecting the environment, reducing costs and producing additional jobs in the management and handling of wastes back into the industrial cycle. A Zero waste strategy may be applied to businesses, communities, industrial sectors, schools and homes.

Benefits proposed by advocates include:

- Saving money. Since waste is a sign of inefficiency, the reduction of waste can reduce costs.
- Faster Progress. A zero waste strategy improves upon production processes and improving environmental prevention strategies which can lead to take larger, more innovative steps.
- Supports sustainability. A zero waste strategy supports all three of the generally accepted goals of sustainability - economic well-being, environmental protection, and social well-being.
- Improved material flows. A zero waste strategy would use far fewer new raw materials and send no waste materials to landfills. Any material waste would either return as reusable or recycled materials or would be suitable for use as compost.

History

1970s: Zero Waste Systems Inc

The term *zero waste* was first used publicly in the name of a company, Zero Waste Systems Inc. (ZWS), which was founded by PhD chemist Paul Palmer in the mid-1970s in Oakland, California. The mission of ZWS was to find new homes for most of the chemicals being excessed by the nascent electronics industry. They soon expanded their services in many other directions. For example, they accepted free of charge, large quantities of new and usable laboratory chemicals which they resold to experimenters, scientists, companies and tinkerers of every description during the 1970s. ZWS arguably had the largest inventory of laboratory chemicals in all of California, which were sold for half price. They also collected all of the solvent produced by the electronics industry called developer/rinse (a mixture of xylene and butyl acetate). This was put into small cans and sold as a lacquer thinner. ZWS collected all the "reflow oil" created by the printed circuit industry, which was filtered and resold into the "downhole" (oil well) industry. ZWS pioneered many other projects.

Because they were the only ones in the world in this business, they achieved an international reputation. Many magazine articles were written about them and several television shows featured them. The California Integrated Waste Management Board produced a slide show featuring ZWS's business and the EPA published a number of studies of their business, calling them an "active waste exchange".

The heir to the ZWS mantle is the Zero Waste Institute (ZWI), also founded by Paul Palmer. Building on the lessons learned from ZWS, the ZWI considers recycling to be no more than an appendage to garbage creation and the garbage industry. ZWI likewise rejects all attempts to reuse garbage or any kind of waste product. Instead, ZWI calls for the redesign of all of the products of industry and commerce, and the processes that produce, sell and make use of them, so that discard never takes place and there is no waste generated needing to be reused or recycled. Discard is seen as the critical step, a commercial and psychological transfer of responsibility which breaks the chain of custody of a product, removes its owner and subjects it to the degradation of garbage management.

The website offers numerous specific examples of ways in which products can be designed so that discard is unnecessary since the lifetime of the product is extended to at least a threshold value of approximately a human lifetime of 100 years. A fully worked out set of principles and analysis is presented, revolving, among other changes, around standardization, modularization and robust design. A theory of design efficiency leading to design effectiveness is presented, which means that once a product is designed to be used in perpetuity, it can be fitted out with robust features, strong materials and special conveniences that could not be afforded in a product designed to be discarded after a single use. That theory is applied to packages as an example.

The ZWI rejects all association with the world of recycling, pointing out that there is no theory of recycling in existence; only a trusting hope that it can be useful.

2002–2003

The movement gained publicity and reached a peak in 1998–2002, and since then has been moving from "theory into action" by focusing on how a "zero waste community" is structured and behaves. The website of the Zero Waste International Alliance has a listing of communities across the globe that have created public policy to promote zero-waste practices. One can check the Eco-Cycle website for examples of how this large nonprofit is leading Boulder County, Colorado on a Zero-Waste path and watch a 6-minute video about the zero-waste big picture. Finally, there is a USA zero-waste organization named the GrassRoots Recycling Network that puts on workshops and conferences about zero-waste activities.

The California Integrated Waste Management Board established a zero waste goal in 2001. The City and County of San Francisco's Department of the Environment established a goal of zero waste in 2002, which led to the City's Mandatory Recycling and

Composting Ordinance in 2009. With its ambitious goal of zero waste and policies, San Francisco reached a record-breaking 80% diversion rate in 2010, the highest diversion rate in any North American city. San Francisco received a perfect score in the waste category in the Siemens US and Canada Green City Index, which named San Francisco the greenest city in North America.

Present Day

The tension between zero waste, viewed as post-discard total recycling of materials only, and zero waste as the reuse of all high level function remains a serious one today. It is probably the defining difference between established recyclers and emerging zero-wasters. A signature example is the difference between smashing a glass bottle (recovering cheap glass) and refilling the bottle (recovering the entire function of the container).

The tension between the literal application of natural processes and the creation of industry-specific more efficient reuse modalities is another tension. Many observers look to nature as an ultimate model for production and innovative materials. Others point out that industrial products are inherently non-natural (such as chemicals and plastics that are mono-molecular) and benefit greatly from industrial methods of reuse, while natural methods requiring degradation and reconstitution are wasteful in that context.

Biodegradable plastic is the most prominent example. One side argues that biodegradation of plastic is wasteful because plastic is expensive and environmentally damaging to make. Whether made of starch or petroleum, the manufacturing process expends all the same materials and energy costs. Factories are built, raw materials are procured, investments are made, machinery is built and used, humans labor and make use of all normal human inputs for education, housing, food etc. Even if the plastic is biodegraded after a single use, all of those costs are lost so it is much more important to design plastic parts for multiple reuse or perpetual lives. The other side argues that keeping plastic out of a dump or the sea is the sole benefit of interest.

Companies moving towards "zero landfill" plants include Subaru, Xerox and Anheuser-Busch.

The movement continues to grow among the youth around the world under the organization Zero Waste Youth, which originated in Brazil and has spread to Argentina, Puerto Rico, Mexico, the United States, and Russia. The organization multiplies with local volunteer ambassadors who lead zero waste gatherings and events to spread the zero waste message.

Packaging Example

Milk can be shipped in many forms. One of the traditional forms was reusable returnable glass milk bottles, often home delivered by a milkman. While some of this continues, other options have recently been more common: one-way gable-top paperboard cartons, one-way aseptic cartons, one-way recyclable glass bottles, one-way milk bags, and others. Each system claims some advantages and also has possible disadvantages. From the zero waste standpoint, the reuse of bottles is beneficial because the material usage per trip can be less than other systems. The primary input (or resource) is silica-sand, which is formed into glass and then into a bottle. The bottle is filled with milk and distributed to the consumer. A reverse logistics system returns the bottles for cleaning, inspection, sanitization, and reuse. Eventually the heavy duty bottle would not be suited for further use and would be recycled. Waste and landfill usage would be minimized. The material waste is primarily the wash water, detergent, transportation, heat, bottle caps, etc. While true zero waste is never achieved, a life cycle assessment can be used to calculate the waste at each phase of each cycle.

Returnable glass milk bottles

Recycling and Rotting (Composting)

It is important to distinguish recycling from Zero Waste.

Some claim that the key component to zero waste is recycling while others reject that notion in favor of reusing high function. The common understanding of recycling is simply that of placing bottles and cans in a recycle bin. The modern version of recycling is more complicated and involves many more elements of financing and government support. For example, a 2007 report by the U.S. Environmental Protection Agency states that the US recycles at a national rate of 33.4% and includes in this figure composted materials. In addition many worldwide commodity industries have been created to handle the materials that are recycled. At the same time, claims of recycling rates have sometimes been exaggerated, for example by the inclusion of soil and organic matter used to cover garbage dumps daily, in the "recycled" column. In states with recycling incentives, there is constant local pressure to pump up the recycling rate figures.

The movement toward recycling has separated itself from the concept of zero waste. One example of this is the computer industry where worldwide millions of PC's are disposed of as electronic waste each year (160 million in 2007). Those computers that enter the recycling stream are broken down into a small amount of raw materials while most merely enter dumps through export to third world countries. Companies are then able to purchase some raw materials, notably steel, copper and glass, reducing the use of new materials. On the other hand, there is an industry, more aligned with the Zero Waste principle of design for long term reuse, that actually repairs computers. It is called the Computer Refurbishing industry and it predates the current campaign to just collect and ship electronics. They have organizations and conferences and have for many years donated computers to schools, clinics and non-profits. Zero Waste planning demands that components be redesigned for effective reuse over long lives leading to even more refurbishing and repair.

There is one seminal example that brings out the difference between Zero Waste and recycling in stark relief. That example, quoted in Getting To Zero Waste, is the software business. Zero Waste is sensitive to the waste of intellectual effort that would be caused by the need to recreate certain basic inventions of software (called objects in software design) as opposed to copying them over and over whenever needed. The waste would occur as the software developers consume resources while solving problems already solved earlier. The application of Zero Waste analysis is straightforward as it recommends conserving human effort. On the other hand, the usual approach of recycling would be to look for some materials that could be found to reuse. The materials on which software is saved (such as paper or diskettes)is of little significance compared to the saving of human effort and if software is saved electronically, there is no media at all. Thus Zero Waste correctly identifies a wasteful behavior to avoid while recycling has no application.

The recycling movement has been embraced by the garbage industry because it serves so well as greenwashing i.e. a way to show that design for garbage creation is acceptable because materials will be kept out of a dump by recycling them. Zero Waste, on the other hand, offers the garbage industry no such screen against public condemnation of waste, and therefore actually threatens the continued need for garbage disposal. For example, in Alameda County, California, garbage dumping is charged a surcharge of $8/ton (as of 2009) which goes entirely for a recycling subsidy but none of which goes for any kind of Zero Waste style designing. Zero Waste has received no support from the garbage industry or politicians under their control except in those cases where it can be claimed to consist solely of more recycling.

Reduce, Refuse And Reuse

Zero waste is poorly supported by the enactment of government laws to enforce the waste hierarchy of refuse, reduce, reuse, recycle and rot (compost). In practice, these laws invariably emphasize destruction and recycling, while the reuse component is marginalized.

A special feature of Zero Waste as a design principle is that it can be applied to any product or process, in any situation or at any level. Thus it applies equally to toxic chemicals as to benign plant matter. It applies to the waste of atmospheric purity by coal burning or the waste of radioactive resources by attempting to designate the excesses of nuclear power plants as "nuclear waste". All processes can be designed to minimize the need for discard, both in their own operations and in the usage or consumption patterns which the design of their products leads to. Recycling, on the other hand, deals only with simple materials.

Zero Waste can even be applied to the waste of human potential by enforced poverty and the denial of educational opportunity. It encompasses redesign for reduced energy wasting in industry or transportation and the wasting of the earth's rainforests. It is a general principle of designing for the efficient use of all resources, however defined.

The recycling movement may be slowly branching out from its solid waste management base to include issues that are similar to the community sustainability movement.

Zero waste on the other hand, is not based in waste management limitations to begin with but requires that we maximize our existing reuse efforts while creating and applying new methods that minimize and eliminate destructive methods like incineration and recycling. Zero Waste strives to ensure that products are designed to be repaired, refurbished, re-manufactured and generally reused.. ("What is Zero Waste?", para 2).

Online web services, like Free Cycle or the reGives Network have risen in popularity over the last decade where locals can give items that they no longer need to others locally in an effort to keep items out of landfills and work toward a zero waste lifestyle.

Significance of Dump Capacity

Many dumps are currently exceeding carrying capacity. This is often, mistakenly used as a justification for moving to Zero Waste. Others counter by pointing out that there are huge tracts of land available throughout the USA and other countries which could be used for dumps. The underlying need to move to a society designed along Zero Waste principles arises from the huge waste of resources that is inherent in poorly made, short-lived articles and production processes. The locus of the most egregious wasting takes place as articles are built and processes are run wastefully. The actual placing of a now useless item in a dump is barely the icing on the cake, in terms of the waste it represents. Poorly conceived proposals, that appear with a dismaying regularity on the Internet, to blithely destroy all garbage as a way to solve the garbage problem, make use of the common delusion that it is the garbage itself which is the problem. These proposals typically claim to convert all or a large portion of existing garbage into oil and sometimes claim to produce so much oil that the world will henceforth have abundant liquid fuels. One such plan, called Anything Into Oil was promoted by Discover Magazine and Fortune Magazine in 2004, even though it absurdly claimed to be able to convert a refrigerator into "light Texas crude" by the application of high pressure steam. Zero Waste analysis, which is long on scientific results and short on spectacular claims, receives no such promotion by the media.

Corporate Initiatives

An example of a company that has demonstrated a change in landfill waste policy is General Motors (GM). GM has confirmed their plans to make approximately half of its 181 plants worldwide "landfill-free" by the end of 2010. Companies like Subaru, Toyota, and Xerox are also producing landfill-free plants. GM is supposed to have about eighty producing plants twenty months. Furthermore, The United States Environmental Protection Agency (EPA) has worked with GM and other companies for decades to minimize the waste through its WasteWise program. The goal for General Motors is finding ways to recycle or reuse more than 90% of materials by: selling scrap materials, adopting reusable parts boxes to replace cardboard, and even recycling used work gloves. The remainder of the scraps might be incinerated to create energy for the plants. Besides being nature friendly, it also saves money by cutting out waste and producing a more efficient production. All these organizations push forth to make our world clean and producing zero waste.

Re-use or Rot of Waste

The waste sent to landfills may be harvested as useful materials, such as in the production of solar energy or natural fertilizer /de-composted manure for crops.

It may also be reused and recycled for something that we can actually use. "The success of General Motors in creating zero-landfill facilities shows that zero-waste goals can be a powerful impetus for manufacturers to reduce their waste and carbon footprint," says Latisha Petteway, a spokesperson for the EPA.

Construction and Deconstruction

Zero Waste is a goal, a process, a way of thinking that profoundly changes our approach to resources and production. Zero Waste is not about recycling and diversion from landfills but about restructuring production and distribution systems to prevent waste from being manufactured in the first place. The materials that are still required in these re-designed, resource-efficient systems will be reused many times as the products that incorporate them are reused. Deconstruction can be described as construction in reverse. It involves carefully taking apart a building to maximize the reuse of materials, thereby reducing waste and conserving resources. Deconstruction can capture materials and some components from the millions of buildings that are existing and that were poorly designed for high level reuse but it is not a favored approach from a Zero Waste point of view. Zero Waste favors the design of buildings as assemblages of high level components, not their creation from rough materials such as lumber, cement or plaster. The details are not worked out yet but to the extent that entire rooms, entire walls, roofs or floors or entire utility systems can be pre-built and installed as completed components, that will be the goal of Zero Waste design. Until buildings are built as components capable of later dismantling, deconstruction is a stop-gap process that the United States can use to minimize the waste of building materials. For now, the largest parts that we are able to save tend to be architectural elements, windows, doors, and metals, many of which are being saved and resold by reuse yards such as Urban Ore in Berkeley, California. The main parts that still need to be crushed are wood flooring, brick walls, and structural timbers. The demolition of traditional buildings has been long done by wrecking ball or bulldozer. Social and political artifacts, such as demolition contractor licenses and required permits that can only be satisfied by destruction and discard (with partial recycling of rubble and steel), render the destruction and disposal costs cheaper than deconstruction. Approximately seventy pounds of the waste is generated for about every square foot of the residential building demolition. It is arguable that this is artificial economics, based on the cultural preference for wastefulness and that Zero Waste designs of dismantlable components will ultimately be the cheapest as well as the most conservative way to reuse buildings. Further discussions of this topic may be found on the ZWI website.

Roper's comments in the paragraph above are either misquoted or wrong concerning wood flooring, structural wood and bricks needing to be crushed. Brick, wood and stone are among the oldest truly recyclable materials used in construction. A historic review of old buildings, barns and bridges clearly shows that brick, stones and timber are reused from older buildings. Some of the oldest structures on the planet are built with materials that were recycled from previous structures. One recent example is the Mayflower Barn at Jordans just north of High Wycombe, UK. The barn is clearly built of reused timbers, possibly sourced from the salvage of the Mayflower ship. It is simply a fact of life that historically materials that could be reused were reused.

In more recent construction, structural timber components, including large timbers, glued laminated beams, floor joists, studs and flooring are some of the most valuable structural components salvaged when a structure is demolished if there is an interest in salvaging. If you need proof, go down to any local construction salvage yard and look at the value of trusses, wood beams, floor joists, studs and flooring. Today they have value when someone saves them.

One of the barriers of reusing structural materials is the bias of building code officials and building departments that discriminate against reusing materials. Codes and building departments require compliance to codes, including the source of materials. Your average contractor cannot just use 100-year-old 2x8 (50x200) salvaged floor joists because the building department requires a graded joist. The contractor then has to find an engineer or wood technologist to verify the material suitability for its use. While the codes technically allow this under alternative method, modern attention to cost usually prevents that option.

Market-based Campaigns

Market-based, legislation-mediated campaigns like Extended Producer Responsibility (EPR) and the Precautionary Principle are among numerous campaigns that have a Zero Waste slogan hung on them by means of claims they will ineluctably lead to policies of Zero Waste. At the moment, there is no evidence that EPR will increase reuse, rather than merely moving discard and disposal into private-sector dumping contracts. The Precautionary Principle is put forward to shift liability for proving new chemicals are safe from the public (acting as guinea pig) to the company introducing them. As such, its relation to Zero Waste is dubious. Likewise, many organizations, cities and counties have embraced a Zero Waste slogan while pressing for none of the key Zero Waste changes. In fact, it is common for many such to simply state that recycling is their entire goal. Many commercial or industrial companies claim to embrace Zero Waste but usually mean no more than a major materials recycling effort, having no bearing on product redesign. Examples include Staples, Home Depot, Toyota, General Motors and computer take-back campaigns. Earlier social justice campaigns have successfully pressured McDonald's to change their meat purchasing practices and Nike to change its labor practices in Southeast Asia. Those were both based on the idea that organized consumers can be active participants in the economy and not just passive subjects. However, the announced and enforced goal of the public campaign is critical. A goal to reduce waste generation or dumping through greater recycling will not achieve a goal of product redesign and so cannot reasonably be called a Zero Waste campaign.

How to Achieve

National and provincial governments often set targets and may provide some funding, but on a practical level, waste management programs (e.g. pickup, dropoff, or containers for recycling and composting) are usually implemented by local governments, possibly with regionally shared facilities.

Reaching the goal of zero waste requires the products of manufacturers and industrial designers to be easily disassembled for recycling and incorporated back into nature or the industrial system; durability and repairability also reduce unnecessary churn in the product lifecycle. Minimizes packaging also solves many problems early in the supply chain. If not mandated by government, choices by retailers and consumers in favor of zero-waste-friendly products can influence production. To prevent material from becoming waste, consumers, businesses, and non-profits must be educated in how to reduce waste and recycle successfully.

Zero Waste Hierarchy

The "Zero Waste Hierarchy" describes a progression of policies and strategies to support the Zero Waste system, from highest and best to lowest use of materials. It is designed to be applicable to all audiences, from policy-makers to industry and the individual. It aims to provide more depth to the internationally recognized 3Rs (Reduce, Reuse, Recycle); to encourage policy, activity and investment at the top of the hierarchy; and to provide a guide for those who wish to develop systems or products that move us closer to Zero Waste. It enhances the Zero Waste definition by providing guidance for planning and a way to evaluate proposed solutions.All over the world, in some form or another, a pollution prevention hierarchy is incorporated into recycling regulations, solid waste management plans, and resource conservation programs. In Canada, a pollution prevention hierarchy otherwise referred to as the Environmental Protection Hierarchy was adopted. This Hierarchy has been incorporated into all recycling regulations within Canada and is embedded within all resource conservation methods which all government mandated waste prevention programs follow. While the intention to incorporate the 4th R (recovery)prior to disposal was good, many organizations focused on this 4th R instead of the top of the hierarchy resulting in costly systems designed to destroy materials instead of systems designed to reduce environmental impact and waste. Because of this, along with other resource destruction systems that have been emerging over the past few decades, Zero Waste Canada along with the Zero Waste International Alliance have adopted the only internationally peer reviewed Zero Waste Hierarchy that focuses on the first 3Rs; Reduce, Reuse and Recycle including Compost.

Zero Waste Jurisdictions

Various governments have declared zero waste as a goal, including:

- California
- Kamikatsu, Tokushima
- Fort Collins, Colorado
- Capannori

An example of network governance approach can be seen in the UK under New Labour who proposed the establishment of regional groupings that brought together the key stakeholders in waste management (local authority representatives, waste industry, government offices etc.) on a voluntary basis. There is a lack of clear government policy on how to meet the targets for diversion from landfill which increases the scope at the regional and local level for governance networks. The overall goal is set by government but the route for how to achieve it is left open, so stakeholders can coordinate and decide how best to reach it.

Zero Waste is a strategy promoted by environmental NGOs but the waste industry is more in favour of the capital intensive option of energy from waste incineration. Research often highlights public support as the first requirement for success. In Taiwan, public opinion was essential in changing the attitude of business, who must transform their material use pattern to become more sustainable for Zero Waste to work.

The latest development in Zero Waste is the city of Masdar in Abu Dhabi which promises to be a Zero Waste city. Innovation and technology is encouraged by government creating an innovation friendly environment without being prescriptive. To be a successful model of sustainable urban development it will also require the involvement and co-operation from all members of society emphasizing the importance of network governance.

Eco-costs Value Ratio

The EVR model is a Life Cycle Assessment based method to analyse consumption patterns, business strategies and design options in terms of eco-efficient value creation. Next to this it is used to compare products and service systems (e.g. benchmarking).

The eco-costs/value ratio (EVR) is an indicator to reveal sustainable and unsustainable consumption patterns of people. The eco-costs is an indicator for the environmental pollution of the products people buy, the value is the price they pay for it in our free market economy. Example: When somebody spends 1000 euro per month on housing (in Europe: EVR approx. 0,3) it is less harmful for the environment than when 1000 euro is spend on diesel (in Europe: EVR approx. 1,0).

The EVR is also relevant for business strategies, because companies are facing the slow but inevitable internalization of environmental costs. At the moment the costs of products don't take into account the environmental damage caused by these products. This "pollution is for free" mentality is less and less accepted by communities.

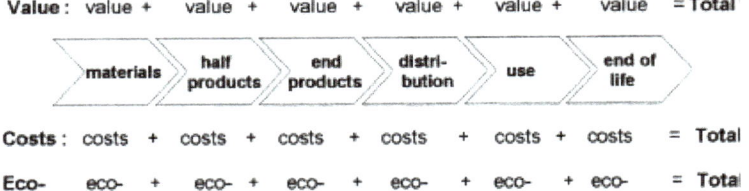

The basic idea of combining the economic and ecological chain: 'the EVR chain',

The EVR makes companies aware of the relative importance of the environmental pollution of their products, and the relative risk they run that future production costs will increase because of this internalization of environmental costs. By using the EVR, companies can make decisions for their product portfolio: abandon products with low value and high environmental costs and stimulate products with high value and low environmental costs.

Background Information

The EVR model has been introduced in 1998 and published in 2000-2004 in the International Journal of LCA, and in the Journal of Cleaner Production. In 2007, and in 2012, the system was updated. The concept of EVR is based on eco-costs. In 2010 a book named "LCA-based assessment of sustainability: the Eco-costs/Value Ratio (EVR)" was published containing the most important articles about the EVR.

Working Principle

The Model

EVR = Eco-costs/value. The basic idea of the EVR model is to link the 'value chain' to the ecological product chain. In the value chain, the added value (in terms of money) and the added costs are determined for each step of the product 'from cradle to grave'. Similarly, the ecological impact of each step in the product chain is expressed in terms of money, the so-called 'eco-costs'. Note that there exists also a Porter chain from the right to the left in Figure, starting with waste and adding value by recycling. In this way the Porter chain becomes circular.

Eco-costs

Eco-costs express the amount of environmental burden of a product on basis of prevention of that burden. They are the marginal prevention costs (money) which should be made to reduce the environmental pollution and materials depletion in our world to a level which is in line with the carrying capacity of our earth.

As such, the eco-costs are virtual costs, since they are not yet integrated in the real life costs of current production chains (Life Cycle Costs). The eco-costs should be regarded as hidden obligations.

For example: for each 1000 kg CO_2 emission, one should invest € 135,- in offshore windmill parks (or other CO_2 reduction systems at that price or less). When this is done consequently, the total CO_2 emissions in the world will be reduced by 65% compared to the emissions in 2008. As a result global warming will stabilise. In short: "the eco-costs of 1000kg CO_2 are € 135,-". Similar calculations can be made on the environmental burden of acidification, eutrification, summer smog, fine dust, eco-toxicity, and the use of metals, fossil fuels and land (nature).

Eco-costs are used in Life Cycle Assessment, LCA, to assess the environmental performance of different materials, processes and End of Life methods.

EVR of Products

The EVR combines eco-cost and value to see whether a product will be successful. The product should have low environmental impact in its lifecycle (low eco-costs) and an attractive value for consumers. The value here is the market value (perceived customer value, also called fair price). First figure below depicts the three dimensions of a product: the value, the costs and the eco-costs.

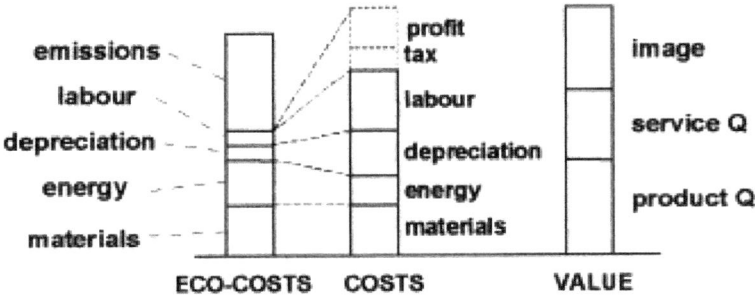

The decomposition of 'virtual eco-costs', costs and value of a product.

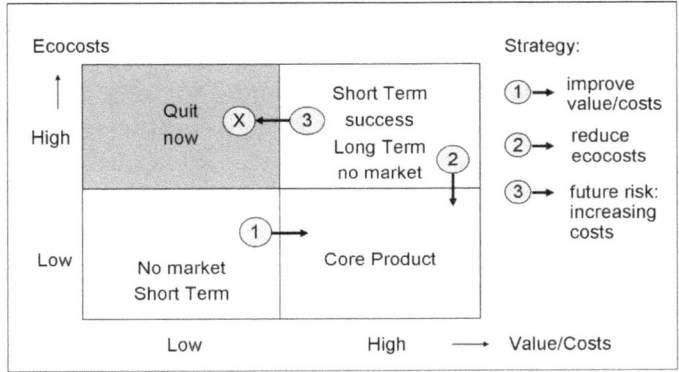

Product portfolio matrix for EVR product strategy of companies.

It is a trend in society that heavy pollution of industry is not accepted anymore by the inhabitants of a country. This results in stricter regulations by countries (e.g. tradable emission rights, enforcement of best available technologies, eco-taxes, etc.). Eco-costs will then become part of the internal production costs. This internalizing of eco-costs might be a threat to a company, but it might also be an opportunity: "When my product has less eco-burden than that of my competitor, my product can withstand stricter regulations of the government. So this characteristic of low eco-costs of my product is a competitive edge." To analyse the short term and the long term market prospects of a product or a product service combination (Product Service System, PSS), each product or PSS can be positioned in the portfolio matrix of Figure . The basic idea of the product portfolio matrix is the notion that a product, service or PSS is characterized by:

- its short term market potential: high value/costs ratio
- its long term market requirement: low eco-costs.

In terms of product strategy, the matrix results in 3 strategic directions:

1. enhance the value/costs ratio of a green design to create a bigger market

2. lower the eco-costs of current successful products to make it fit for future markets

3. abandon products with a low value/ costs ratio (not much profit, small market) and high eco-costs

For many 'green designs', the usual problem is that they have a low current value/costs ratio. In most of the cases the production costs are higher than the production costs of the classic solution, in some cases even the (perceived) quality is poor. There are two ways to do something about it:

a. enhance the (perceived) quality of the product

b. attach to the product a service (create a PSS) in a way that the value of the bundle of the product and the service is more than the value of its components.

For a product which has a good present value/costs ratio, but high eco-costs, the product and the production process have to be redesigned to lower the eco-costs. This road towards sustainability is often far more promising than the strategy of enhancing the value/costs ratio of a green design.

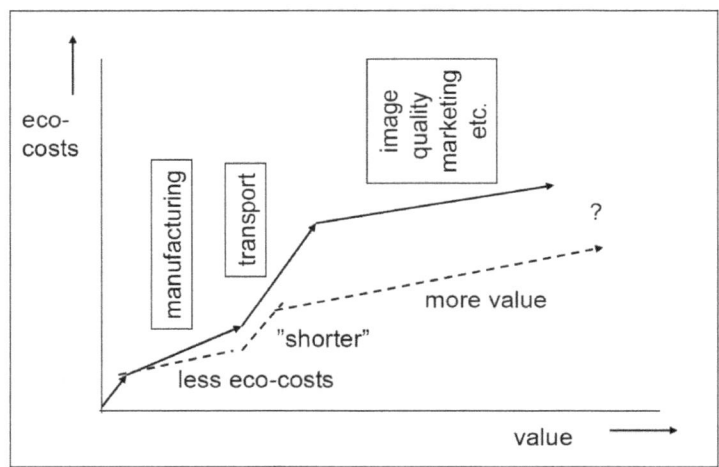

Design strategies to enhance the EVR of a product.

The reason is that the economies of scale for production and distribution are available and that the new product is marketed to an existing client base which is used to the brand name, the quality standards, the service system, etc.

The most common fear of business managers is that their new green products end up with a deteriorated value/costs ratio, and hence will have a cumbersome position in the market. The stability of the governmental policy plays an important role here. When governmental regulations which level the playing field are postponed or even abandoned, proactive companies with sound product strategies are harmed. This can cause severe damage to the transition process and may lead to reluctance of players to move proactively in the future.

The most successful design options are depicted in Figure. The best design strategy is:

- to increase value where value is high
- to decrease the eco-costs where the eco-costs are high

Use

EVR & De-linking

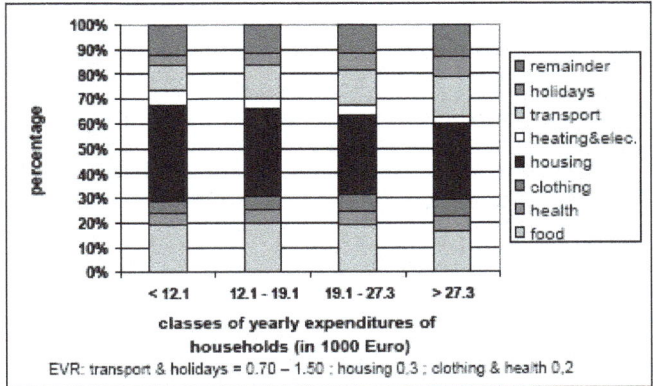

The consumer's side: preference of expenditures in Dutch households.

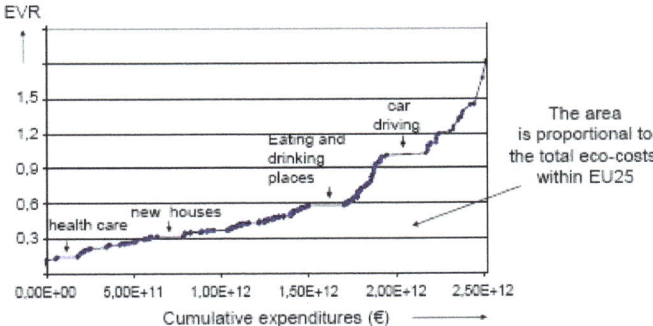

The EVR and the total expenditures of all consumers in the EU25 (from EIPRO)

In economics, de-linking (also known as decoupling) is often used in the context of economic production and environmental quality. In this context, it refers to the ability of an economy to grow without corresponding increases in environmental pressure. In many economies increasing production (GDP) would involve increased pressure on the environment. An economy that is able to sustain GDP growth, without also experiencing a worsening of environmental conditions, is said to be de-linked.

There is a consumer's side of the de-linking of economy and ecology. Under the assumption that most of the households spend in their life what they earn in their life, the total EVR of the spending of households is the key towards sustainability. Only when this total EVR of the spending gets lower, the eco-costs related to the total spending can be reduced even at a higher level of spending. There are two ways of achieving this:

- At the production side: the improvement of eco-efficiency ('lowering EVR') of products and services by the industry

- At the consumer's side: the change of lifestyle of customers in the direction of 'low EVR' products.

At the production side, our society is heading in the right direction: gradually, industrial production is achieving higher levels of the value/costs ratio and is at the same time becoming cleaner. At the consumer's side, however, our society is suffering from the fact that the consumers preferences are heading in the wrong direction: towards products and services with an unfavourable EVR (like driving in SUVs, more kilometres, intercontinental flights for holidays). These unfavourable preferences can be concluded from Figure.

Figure shows that people in the Netherlands (and probably in the other EC countries as well) spend relatively more money on cars and holidays when they have more money available. Other studies show that people tend to have intercontinental holidays at the moment they can afford it. This shift in consumer spending will become a big problem in the near future, since the EVR of e.g. housing and health care is much lower than the EVR of transport and (inter)continental holidays by plane. Figure shows the EVR (= ecocosts/price) on the Y/axis as a function of the cumulative expenditures of all products and services of all citizens in the EU 25 on the X-axis. The data is from the EIPRO study of the European Commission (EIPRO = environmental impact of products).

The area underneath the curve is proportional to the total eco-costs of the EU25. Basically there are two strategies to reduce the area under the curve: - ask industry to reduce the eco-costs of their products (this will shift the curve downward) - try to reduce expenditures of consumers in high end of the curve, and let them spend this money at the low end of the curve (this will shift the middle part of the curve to the right). The question is now how designers and engineers can contribute to this required shift towards sustainability and what this means to product portfolio strategies of companies. The solution is Eco-efficient Value Creation.

Eco-efficient Value Creation

The way towards sustainability requires a double aim in product innovation:

- lower eco-costs, and at the same time
- higher value (a higher market price).

We call this: Eco-efficient Value Creation. The reason we need value creation for eco-efficient products is threefold:

1. the higher price in the market is required to cover the higher production cost of green products (note that a higher price is only accepted by the consumer when the perceived value is higher, otherwise the consumer will not buy the product)

2. the higher price prevents the rebound effect

3. lowering the EVR appears the key to a sustainable development at the level of countries

Below, an example of eco-efficient value creation is given, which is the introduction of the Lexus RX 400h in the USA:

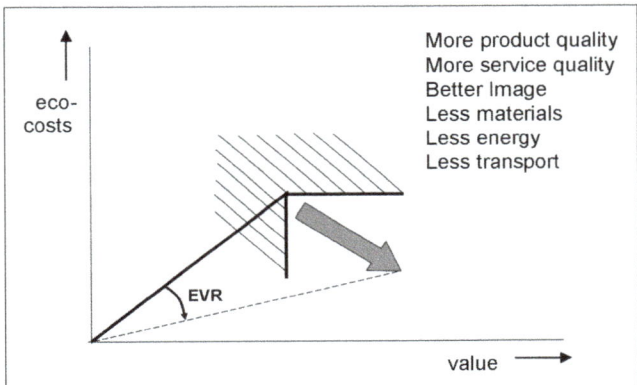

The double objectvive for design & engineering: less eco-costs, more value.

- the customer value has increased, by emphasising its combined power and comfort (from the advertisement in the US: "While it may have a V6 engine under the hood, the extra boost from the electric-drive motor gives the vehicle the acceleration power of a V8 and the noise levels in Lexus hybrid vehicles have been reduced even more")
- the eco-costs of driving are lower, since its excellent overall fuel economy

Note that the acceleration of a car is an interesting issue in terms of value. High acceleration is associated with expensive sports cars (Porsche, Ferrari). But people who buy these fast cars hardly use it. For these people acceleration is more part of the image of the product than it is part of the product qualities they use on a daily basis. So reducing the acceleration is the wrong strategy: it eliminates the extra value, and it hardly reduces the overall eco-costs in practice.

Environmental Benchmarking in LCA

Life Cycle Assessment (LCA) is the generally accepted method to compare two (or more) alternative products or services. A prerequisite for such a comparison is that the functionality ('functional unit') and the quality of the alternatives are the same (you cannot compare apples and oranges in the classical LCA). In cases of product design and architecture, however, this prerequisite seems to be a fundamental flaw in the application of LCA: the designer or architect is aiming at a better quality (in the broad sense of the word: including intangible aspects like beauty and image), so the new design never has the same quality. In some cases the functionality of the design is not the same, since the design solution is limited by a maximum budget, in some cases the functionality is the same, but the higher quality results in a higher price. In all these cases a single indicator in LCA (like the eco-costs) is not suitable for environmental benchmarking. In these cases however, it does make sense to compare the design alternatives on the basis of the eco-costs/value ratio (EVR), where the value is the perceived customer value (the fair price).

Example 1. Different types of armchairs differ in terms of comfort, aesthetics, etc. rather than in terms of functionality. A classical LCA (with a single indicator like eco-costs, carbon footprint, etc.) does not make sense here. Selection on the basis of EVR, however, is the key to a sustainable consumption pattern. The chair with the lowest EVR is the best solution in terms of sustainability.

Example 2. In LCA, the comparison of a new building and a renovated building is in the majority of cases not possible, since, in practice, both solutions differ in almost all quality aspects (tangible as well as intangible). However, the solution with lowest EVR is the best in terms of sustainable consumption.

Note that the renovated building is the best solution in most of the cases, because it has the lowest EVR in the production phase. However, in some cases the renovated building is not the best solution, because of unfavourable energy consumption (high EVR) in the use phase.

Ecological Footprint

The ecological footprint measures human demand on nature, i.e., the quantity of nature it takes to support people or an economy. It tracks this demand through an ecological accounting system. The accounts contrast the biologically productive area people use for their consumption to the biologically productive area available within a region or the world (biocapacity). In short, it is a measure of human impact on Earth's ecosystem and reveals the dependence of the human economy on natural capital.

The ecological footprint is defined as the biologically productive area needed to provide for everything people use: fruits and vegetables, fish, wood, fibers, absorption of carbon dioxide from fossil fuel use, and space for buildings and roads. Biocapacity is the productive area that can regenerate what people demand from nature.

Methods and Techniques of Industrial Ecology

The natural resources of the Earth are finite.

Footprint and biocapacity can be compared at the individual, regional, national or global scale. Both footprint and biocapacity change every year with number of people, per person consumption, efficiency of production, and productivity of ecosystems. At a global scale, footprint assessments show how big humanity's demand is compared to what planet Earth can renew. Global Footprint Network calculates the ecological footprint from UN and other data for the world as a whole and for over 200 nations. They estimate that as of 2013, humanity has been using natural capital 1.6 times as fast as nature can renew it.,

Ecological footprint analysis is widely used around the Earth in support of sustainability assessments. It can be used to measure and manage the use of resources throughout the economy and explore the sustainability of individual lifestyles, goods and services, organizations, industry sectors, neighborhoods, cities, regions and nations. Since 2006, a first set of ecological footprint standards exist that detail both communication and calculation procedures. The latest version are the updated standards from 2009

Footprint Measurements

In 2013, the Global Footprint Network estimated the global ecological footprint as 1.6 planet Earths. This means that, according to their calculations, the planet's ecological services were being used 1.6 times faster than they were being renewed.

Ecological footprints can be calculated at any scale: for an activity, a person, a community, a city, a region, a nation or humanity as a whole. Cities, due to population concentration, have large ecological footprints and have become ground zero for footprint reduction.

Global Footprints: Currently there is no fixed way to measure global footprints, and any attempts to describe the capacity of an ecosystem in a single number is a massive simplification of thousands of key renewable resources, which are not used or replenished at the same rate. However, there has been some convergence of metrics and standards since 2006.

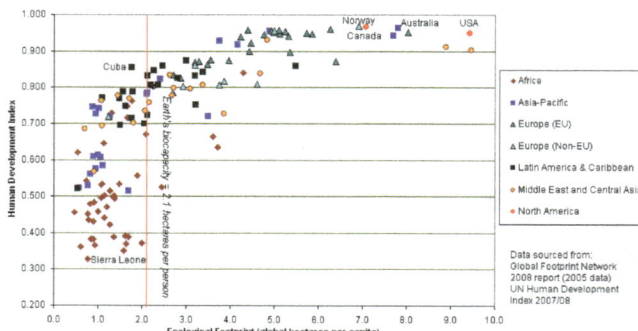

Ecological footprint for different nations compared to their Human Development Index.

City Ecological Footprints: are being measured. There are two types of measurements in use. The first measures ecosystem displacement which is defined as City Area minus remaining green spaces. This is an area measurement that does not include human or other biological activity. The Second attempts to quantify surviving ecosystem health. Specifically, it attempts to quantify both area and biological health of ecosystems surviving inside city areas such as nature reserves, parks, other green spaces. City footprints are being calculated and ranked with city ecological indexes.

Overview

The first academic publication about ecological footprints was by William Rees in 1992. The ecological footprint concept and calculation method was developed as the PhD dissertation of Mathis Wackernagel, under Rees' supervision at the University of British Columbia in Vancouver, Canada, from 1990–1994. Originally, Wackernagel and Rees called the concept "appropriated carrying capacity". To make the idea more accessible, Rees came up with the term "ecological footprint", inspired by a computer technician who praised his new computer's "small footprint on the desk". In early 1996, Wackernagel and Rees published the book *Our Ecological Footprint: Reducing Human Impact on the Earth* with illustrations by Phil Testemale.

Footprint values at the end of a survey are categorized for Carbon, Food, Housing, and Goods and Services as well as the total footprint number of Earths needed to sustain the world's population at that level of consumption. This approach can also be applied to an activity such as the manufacturing of a product or driving of a car. This resource accounting is similar to life-cycle analysis wherein the consumption of energy, biomass (food, fiber), building material, water and other resources are converted into a normalized measure of land area called global hectares (gha).

Per capita ecological footprint (EF), or ecological footprint analysis (EFA), is a means of comparing consumption and lifestyles, and checking this against nature's ability to provide for this consumption. The tool can inform policy by examining to what extent a nation uses more (or less) than is available within its territory, or to what extent the nation's lifestyle would be replicable worldwide. The footprint can also be a useful tool to educate people about carrying capacity and overconsumption, with the aim of altering personal behavior. Ecological footprints may be used to argue that many current lifestyles are not sustainable. Such a global comparison also clearly shows the inequalities of resource use on this planet at the beginning of the twenty-first century.

In 2007, the average biologically productive area per person worldwide was approximately 1.8 global hectares (gha) per capita. The U.S. footprint per capita was 9.0 gha, and that of Switzerland was 5.6 gha, while China's was 1.8 gha. The WWF claims that the human footprint has exceeded the biocapacity (the available supply of natural resources) of the planet by 20%. Wackernagel and Rees originally estimated that the available biological capacity for the 6 billion people on Earth at that time was about 1.3 hectares per person, which is smaller than the 1.8 global hectares published for 2006, because the initial studies neither used global hectares nor included bioproductive marine areas.

Methodology

The ecological footprint accounting method at the national level is described in the l Footprint Atlas 2010 or in greater detail in the Calculation Methodology for the National Footprint Accounts. The National Accounts Review Committee has also published a research agenda on how the method will be improved.

In 2003, Jason Venetoulis, Carl Mas, Christopher Gaudet, Dahlia Chazan, and John Talberth developed Footprint 2., which offers a series of theoretical and methodological improvements to the standard footprint approach. The four primary improvements were that they included the entire surface of the Earth in biocapacity estimates, allocated space for other (i.e., non-human) species, updated the basis of equivalence factors from agricultural land to net primary productivity (NPP), and refined the carbon component of the footprint based on the latest global carbon models.

Studies in the United Kingdom

The UK's average ecological footprint is 5.45 global hectares per capita (gha) with variations between regions ranging from 4.80 gha (Wales) to 5.56 gha (East England).

Two recent studies have examined relatively low-impact small communities. BedZED, a 96-home mixed-income housing development in South London, was designed by Bill Dunster Architects and sustainability consultants BioRegional for the Peabody Trust. Despite being populated by relatively "mainstream" home-buyers, BedZED was found to have a footprint of 3.20 gha due to on-site renewable energy production, energy-efficient architecture, and an extensive green lifestyles program that included on-site London's first carsharing club. The report did not measure the added footprint of the 15,000 visitors who have toured BedZED since its completion in 2002. Findhorn Ecovillage, a rural intentional community in Moray, Scotland, had a total footprint of 2.56 gha, including both the many guests and visitors who travel to the community to undertake residential courses there and the nearby campus of Cluny Hill College. However, the residents alone have a footprint of 2.71 gha, a little over half the UK national average and one of the lowest ecological footprints of any community measured so far in the industrialized world. Keveral Farm, an organic farming community in Cornwall, was found to have a footprint of 2.4 gha, though with substantial differences in footprints among community members.

Critiques

Early criticism was published by van den Bergh and Verbruggen in 1999, which was updated in 2014. Another criticism was published in 2008. A more complete review commissioned by the Directorate-General for the Environment (European Commission) was published in June 2008.

A recent critique of the concept is due to Blomqvist et al., 2013a, with a reply from Rees and Wackernagel, 2013, and a rejoinder by Blomqvist et al., 2013b.

An additional strand of critique is due to Giampietro and Saltelli (2014a), with a reply from Goldfinger et al., 2014, a rejoinder by Giampietro and Saltelli (2014a), and additional comments from van den Bergh and Grazi (2015).

A number of countries have engaged in research collaborations to test the validity of the method. This includes Switzerland, Germany, United Arab Emirates, and Belgium.

Grazi et al. (2007) have performed a systematic comparison of the ecological footprint method with spatial welfare analysis that includes environmental externalities, agglomeration effects and trade advantages. They find that the two methods can lead to very distinct, and even opposite, rankings of different spatial patterns of economic activity. However this should not be surprising, since the two methods address different research questions.

Calculating the ecological footprint for densely populated areas, such as a city or small country with a comparatively large population — e.g. New York and Singapore respectively — may lead to the perception of these populations as "parasitic". This is because these communities have little intrinsic biocapacity, and instead must rely upon large *hinterlands*. Critics argue that this is a dubious characterization since mechanized rural farmers in developed nations may easily consume more resources than urban inhabitants, due to transportation requirements and the unavailability of economies of scale. Furthermore, such moral conclusions seem to be an argument for autarky. Some even take this train of thought a step further, claiming that the Footprint denies the benefits of trade. Therefore, the critics argue that the Footprint can only be applied globally.

The method seems to reward the replacement of original ecosystems with high-productivity agricultural monocultures by assigning a higher biocapacity to such regions. For example, replacing ancient woodlands or tropical forests with monoculture forests or plantations may improve the ecological footprint. Similarly, if organic farming yields were lower than those of conventional methods, this could result in the former being "penalized" with a larger ecological footprint. Of course, this insight, while valid, stems from the idea of using the footprint as one's only metric. If the use of ecological footprints are complemented with other indicators, such as one for biodiversity, the problem could maybe be solved. Indeed, WWF's Living Planet Report complements the biennial Footprint calculations with the Living Planet Index of biodiversity. Manfred Lenzen and Shauna Murray have created a modified Ecological Footprint that takes biodiversity into account for use in Australia.

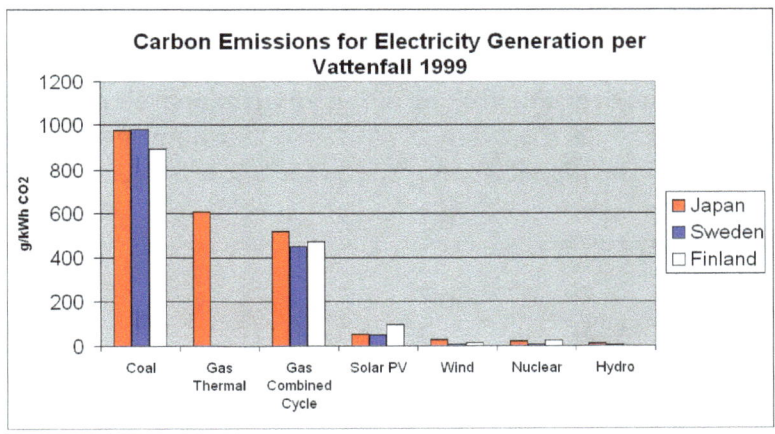

Although the ecological footprint model prior to 2008 treated nuclear power in the same manner as coal power, the actual real world effects of the two are radically different. A life cycle analysis centered on the Swedish Forsmark Nuclear Power Plant estimated carbon dioxide emissions at 3.10 g/kWh and 5.05 g/kWh in 2002 for the Torness Nuclear Power Station. This compares to 11 g/kWh for hydroelectric power, 950 g/kWh for installed coal, 900 g/kWh for oil and 600 g/kWh for natural gas generation in the United States in 1999. Figures released by Mark Hertsgaard, however, show that because of the delays in building nuclear plants and the costs involved, investments in energy efficiency and renewable energies have seven times the return on investment of investments in nuclear energy.

The Swedish utility Vattenfall did a study of full life-cycle greenhouse-gas emissions of energy sources the utility uses to produce electricity, namely: Nuclear, Hydro, Coal, Gas, Solar Cell, Peat and Wind. The net result of the study was that nuclear power produced 3.3 grams of carbon dioxide per KW-Hr of produced power. This compares to

400 for natural gas and 700 for coal (according to this study). The study also concluded that nuclear power produced the smallest amount of CO_2 of any of their electricity sources.

Claims exist that the problems of nuclear waste do not come anywhere close to approaching the problems of fossil fuel waste. A 2004 article from the BBC states: "The World Health Organization (WHO) says 3 million people are killed worldwide by outdoor air pollution annually from vehicles and industrial emissions, and 1.6 million indoors through using solid fuel." In the U.S. alone, fossil fuel waste kills 20,000 people each year. A coal power plant releases 100 times as much radiation as a nuclear power plant of the same wattage. It is estimated that during 1982, US coal burning released 155 times as much radioactivity into the atmosphere as the Three Mile Island incident. In addition, fossil fuel waste causes global warming, which leads to increased deaths from hurricanes, flooding, and other weather events. The World Nuclear Association provides a comparison of deaths due to accidents among different forms of energy production. In their comparison, deaths per TW-yr of electricity produced (in UK and USA) from 1970 to 1992 are quoted as 885 for hydropower, 342 for coal, 85 for natural gas, and 8 for nuclear.

The Western Australian government State of the Environment Report included an Ecological Footprint measure for the average Western Australian seven times the average footprint per person on the planet in 2007, a total of about 15 hectares.

Footprint by Country

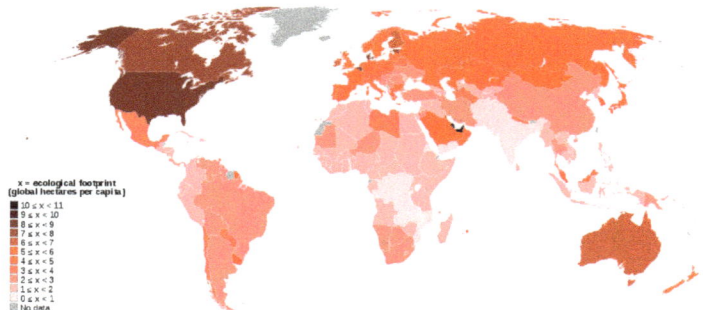

World map of countries by ecological footprint (data of 2007).

The world-average ecological footprint in 2013 was 2.8 global hectares per person. The average per country ranges from over 10 to under 1 global hectares per person. There is also a high variation within countries, based on individual lifestyle and economic possibilities.

The GHG footprint or the more narrow carbon footprint are a component of the ecological footprint. Often, when only the carbon footprint is reported, it is expressed in weight of CO2 (or CO2e representing GHG warming potential (GGWP)), but it can also be expressed in land areas like ecological footprints. Both can be applied to products, people or whole societies.

Implications

the average world citizen has an eco-footprint of about 2.7 global average hectares while there are only 2.1 global hectare of bioproductive land and water per capita on earth. This means that humanity has already overshot global biocapacity by 30% and now lives unsustainabily by depleting stocks of "natural capital"

Energy Accounting

Energy accounting is a system used to measure, analyze and report the energy consumption of different activities on a regular basis. It is done to improve energy efficiency, and to monitor the environment impact of energy consumption.

Energy Management

Thermal energy is the amount of random molecular kinetic energy.

Energy accounting is a system used in energy management systems to measure and analyze energy consumption to improve energy efficiency within an organization. Organisations such as Intel corporation use these systems to track energy usage.

Various energy transformations are possible. An energy balance can be used to track energy through a system. This becomes a useful tool for determining resource use and environmental impacts. How much energy is needed at each point in a system is measured, as well as the form of that energy. An accounting system keeps track of energy in, energy out, and non-useful energy versus work done, and transformations within a system. Sometimes, non-useful work is what is often responsible for environmental problems.

Energy Balance

Energy returned on energy invested (EROEI) is the ratio of energy delivered by an energy technology to the energy invested to set up the technology.

Rebound Effect (Conservation)

The rebound effect (or take-back effect, RE) is the reduction in expected gains from new technologies that increase the efficiency of resource use, because of behavioral or other systemic responses. These responses usually tend to offset the beneficial effects of the new technology or other measures taken. While the literature on the rebound effect generally focuses on the effect of technological improvements on energy consumption, the theory can also be applied to the use of any natural resource or other input, such as labor. The rebound effect is generally expressed as a ratio of the lost benefit compared to the expected environmental benefit when holding consumption constant.

For instance, if a 5% improvement in vehicle fuel efficiency results in only a 2% drop in fuel use, there is a 60% rebound effect (since $(5-2)/5 = 60\%$). The 'missing' 3% might have been consumed by driving faster or further than before.

The existence of the rebound effect is uncontroversial. However, debate continues as to the magnitude and impact of the effect in real world situations. Depending on the magnitude of the rebound effect, there are five different rebound effect types:

1. Super conservation (RE < 0): the actual resource savings are higher than expected savings – the rebound effect is negative. This occurs if the increase in efficiency reduces costs.

2. Zero rebound (RE = 0): The actual resource savings are equal to expected savings – the rebound effect is zero.

3. Partial rebound (0 < RE < 1): The actual resource savings are less than expected savings – the rebound effect is between 0% and 100%. This is sometimes known as 'take-back', and is the most common result of empirical studies on individual markets.

4. Full rebound (RE = 1): The actual resource savings are equal to the increase in usage – the rebound effect is at 100%.

5. Backfire (RE > 1): The actual resource savings are negative because usage increased beyond potential savings – the rebound effect is higher than 100%. This situation is commonly known as the Jevons paradox.

In order to avoid the rebound effect, environmental economists have suggested that any cost savings from efficiency gains be taxed in order to keep the cost of use the same.

History

The rebound effect was first described by William Stanley Jevons in his 1865 book *The*

Coal Question, where he observed that the invention in Britain of a more efficient steam engine meant that the use of coal became economically viable for many new uses. This ultimately led to increased coal demand and much increased coal consumption, even as the amount of coal required for any particular use fell. According to Jevons, "It is a confusion of ideas to suppose that the economical use of fuel is equivalent to diminished consumption. The very contrary is the truth."

However, most contemporary authors credit Daniel Khazzoom for the re-emergence of the rebound effect in the research literature. Although Khazzoom did not use the term, he raised the idea that there is a less than one-to-one correlation between gains in energy efficiency and reductions in energy use, because of a change in the 'price content' of energy in the provision of the final consumer product. His study was based on energy efficiency gains in home appliances, but the principle applies throughout the economy. A commonly studied example is that of a more fuel-efficient car. As each kilometre of travel becomes cheaper, there will be an increase in driving speed and/or kilometres driven, as long as the price elasticity of demand for car travel is not zero. Other examples might include the growth in garden lighting after the introduction of energy-saving compact fluorescent lamps or the increasing size of houses driven partly by higher fuel efficiency in home heating technologies. If the rebound effect is larger than 100%, all gains from the increased fuel efficiency would be wiped out by increases in demand (the Jevons paradox).

Khazzoom's thesis was criticized heavily by Michael Grubb and Amory Lovins who dismissed any disconnection between energy efficiency improvements in an individual market, and an economy-wide reduction in energy consumption. Developing Khazzoom's idea further, and prompting heated debate in the Energy Policy journal at that time, Len Brookes wrote of the fallacies in the energy-efficiency solution to greenhouse gas emissions. His analysis showed that any economically justified improvements in energy efficiency would in fact stimulate economic growth and increase total energy use. For improvements in energy efficiency to contribute to a reduction in economy-wide energy consumption, the improvement must come at a greater economic cost. Commenting in regard to energy efficiency advocates, he concludes that, "the present high profile of the topic seems to owe more to the current tide of green fervor than to sober consideration of the facts, and the validity and cost of solutions."

Khazzoom-Brookes Postulate

In 1992, economist Harry Saunders coined the term "Khazzoom-Brookes postulate" to describe the idea that energy efficiency gains paradoxically result in increases in energy use (the modern day equivalent of the Jevons paradox). He modeled energy efficiency gains using a variety of neo-classical growth models, and showed that the postulate is true over a wide range of assumptions. In the conclusion of his paper, Saunders stated that:

In the absence of efficiency gains, energy use will grow in lock step with economic growth (energy intensity will stay fixed) when energy prices are fixed. ... Energy efficiency gains can increase energy consumption by two means: by making energy appear effectively cheaper than other inputs; and by increasing economic growth, which pulls up energy use. ... These results, while by no means proving the Khazzoom-Brookes postulate, call for prudent energy analysts and policy makers to pause a long moment before dismissing it.

This work provided a theoretical grounding for empirical studies and played an important role in framing the problem of the rebound effect. It also reinforced an emerging ideological divide between energy economists on the extent of the yet to be named effect. The two tightly held positions are:

- Technological improvements in energy efficiency enable economic growth that was otherwise impossible without the improvement; as such, energy efficiency improvements will usually back-fire in the long term.

- Technological improvements in energy efficiency may result in a small take-back. However, even in the long term, energy efficiency improvements usually result in large overall energy savings.

Even though many studies have been undertaken in this area, neither position has yet claimed a consensus view in the academic literature. Recent studies have demonstrated that direct rebound effects are significant (about 30% for energy), but that there is not enough information about indirect effects to know whether or how often back-fire occurs. Economists tend to the first position, but most governments, businesses, and environmental groups adhere to the second. Governments and environmental groups often advocate further research into fuel efficiency and radical increases in the efficient use of energy as the primary means for reducing energy use and reducing greenhouse gas emissions (to alleviate the impacts of climate change). However, if the first position more accurately reflects economic reality, current efforts to invent fuel-efficient technologies may not much reduce energy use, and may in fact paradoxically increase oil and coal consumption, and greenhouse gas emissions, over the long run.

Types of Effects

The full rebound effect can be distinguished into three different economic reactions to technological changes:

1. Direct rebound effect: An increase in consumption of a good is caused by the lower cost of use. This is caused by the substitution effect.

2. Indirect rebound effect: The lower cost of a service enables increased household consumption of other goods and services. For example, the savings from a more efficient cooling system may be put into another luxury good. This is caused by the income effect.

3. Economy wide effect: The fall in service cost reduces the price of other goods, creates new production possibilities and increases economic growth.

In the example of improved vehicle fuel efficiency, the direct effect would be the increased fuel use from more driving as driving becomes cheaper. The indirect effect would incorporate the increased consumption of other goods enabled by household cost savings from increased fuel efficiency. Since consumption of other goods increase, the embodied fuel used in the production of those goods would increase as well. Finally, the economy wide effect would include the long-term effect of the increase in vehicle fuel efficiency on production and consumption possibilities throughout the economy, including any effects on economic growth rates.

Direct and Indirect Effects

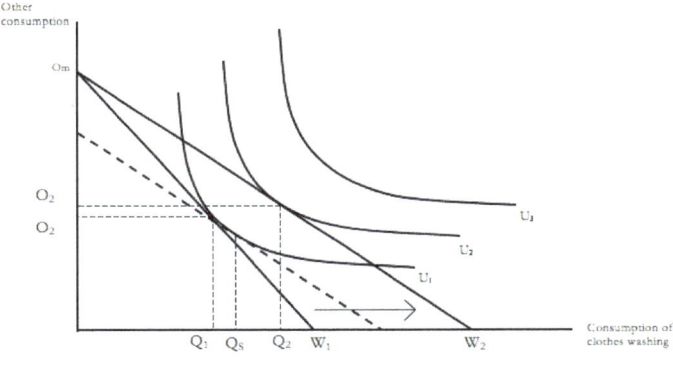

Direct and Indirect Effects

For cost reducing resource efficiency, distinguishing between direct and indirect effects is shown in Figure. The horizontal axis shows units of consumption of the targets good (which could be for example clothes washing, and measured in terms of kilograms of clean clothes) with consumption of all other goods and services on the vertical axis. An economical technology change that enables each unit of washing to be produced with less electricity results in a reduction of the price per unit of washing. This shifts the household budget line rightwards. The result is a substitution effect because of the decreased relative price, but also an income effect due to the increased real income. The substitution effect increases consumption of washing from Q1 to QS, and the income effect from QS to Q2. The total increase in consumption of washing from Q1 to Q2 and the resulting increase in electricity consumption is the direct effect. The indirect effect comprises the increase in other consumption, from O1 to O2. The scale of each of these effects depends on the elasticity of demand for each of the goods, and the embodied resource or externality associated with each good. Indirect effects are difficult to measure empirically. In the manufacturing sector, it has been estimated that there is about a 24% rebound effect due to increases in fuel efficiency. A parallel effect will happen for cost saving efficient technologies for producers, where output and substitution effects will occur.

The rebound effect can increase the difficulty of projecting the reduction in greenhouse emissions from an improvement in energy efficiency. Estimation of the scale of direct effects on residential electricity, heating and motor fuel consumption has been common motivation for research of rebound effects. Evaluation and econometric methods are the two approaches generally employed in estimating the size of this effect. Evaluation methods rely on quasi-experimental studies and measure the before and after changes to energy consumption from the implementation of energy efficient technology, while econometric methods utilize elasticity estimates to forecast the likely effects from changes in the effective price of energy services.

Research has found that in developed countries, the direct rebound effect is usually small to moderate, ranging from roughly 5% to 40% in residential space heating and cooling. Some of the direct rebound effect can be attributed to consumers who were previously unable to use a service. However, the rebound effect may be more significant in the context of the undeveloped markets in developing economies.

Indirect Effects from Conservation

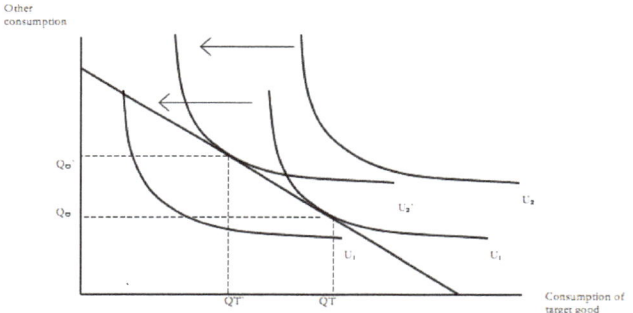

Change in preferences of a household revealing indirect effects from conservation

For conservation measures, indirect effects closely approximate the total economy wide effect. Conservation measures constitute a change in consumption patterns away from particular targeted goods towards other goods. Figure above shows that a change in preference of a household results in a new consumption pattern that has less of the target good (QT to QT`), and more of all other goods (QO to QO`). The resource consumption or externalities embodied in this other consumption is the indirect effect.

Although a persuasive view has prevailed that indirect effects with respect to energy and greenhouse emissions should be very small due to energy directly comprising only a small component of household expenditure, this view is gradually being eroded. Many recent studies based on life-cycle analysis show the energy consumed indirectly by households is often higher than consumed directly through electricity, gas, and motor fuel, and is a growing proportion. This is evident in the results of recent studies that indicate indirect effects from household conservation can range from 10% to 200% depending on the scenario, with higher indirect rebounds from diet changes aiming to reduce food miles.

Economy Wide Effects

Even if the direct and indirect rebound effects add up to less than 100%, technological improvements that increase efficiency may still result in economy wide effects that results in increased resource use for the economy as a whole. In particular, this would happen if resource efficiency enables an expansion of production in the economy, and an increase in the rate of economic growth. For example, for the case of energy use, more efficient technology is equivalent to a lower price for energy resources. It is well known that changes in energy costs have a large impact on economic growth rates. In the 1970s sharp increases in petroleum prices led to stagflation (recession and inflation) in the developed countries, whereas in the 1990s lower petroleum prices contributed to higher economic growth. An improvement in energy efficiency has the same effect as lower fuel prices, and leads to faster economic growth. Economists generally believe that especially for the case of energy use, more efficient technologies will lead to increased use, because of this growth effect.

To model the scale of this effect, economists use computational general equilibrium (CGE) models. While CGE methodology is by no means perfect, results indicate that economy wide rebound effects are likely to be very high, with estimates above 100% rather common. One simple CGE model has been made available online for use by economists.

Income Level Variation

Research has shown that the direct rebound effects for energy services is lower at high income levels, due to less price sensitivity. Studies have found that own-price elasticity of gas consumption by UK households was two times greater for households in the lowest income decile when compared to the highest decile. Studies have also observed higher rebounds in low-income houses for improvements in heating technology. Evaluation methods have also been used to assess the scale of rebound effects from efficient heating installations in lower income homes in the United Kingdom. This research found that direct effects are close to 100% in many cases. High income households in developed countries are likely to set the temperature at the optimum comfort level, regardless of the cost – therefore any cost reduction does not result in increased heating, for it was already optimal. But low-income households are more price sensitive, and have made thermal sacrifices due to the cost of heating. In this case, a high direct rebound is likely. This analogy can be extended to most household energy consumption.

The size of the rebound effect is likely to be higher in developing countries according to macro-level assessments and case studies. One case study was undertaken in rural India to evaluate the impact of an alternative energy scheme. Households were given solar powered lighting in an attempt to reduce the use of kerosene for lighting to zero except for seasons with insufficient sunshine. The scheme was also designed to encourage a future willingness to pay for efficient lighting. The results were surprising, with high direct rebounds between 50 and 80%, and total direct and indirect rebound above 100%. Because the new lighting source was essentially zero cost, operating hours for lighting went up from an average of 2 to 6 per day, with new lighting consisting of a combination of both the no-cost solar lamps and also kerosene lamps. Also, more cooking was undertaken which enabled an increased trade of food with neighboring villages.

Rebounds with Respect to Time

The individual opportunity of cost is an often overlooked cause of rebound effect. Time is being substituted with an increase in demand for a service. Research articles often examine increasingly convenient and more rapid modes of transportation to determine the rebound effect in energy demand. Because time cost forms a major part of the total cost of commuter transport, rapid modes will reduce real costs, but will also encourage longer commuting distances which will in turn increase energy consumption. While important, it is almost impossible to estimate empirically the scale of such effects due to the subjective nature of the value of time. Time saved can either be used towards additional work or leisure which may have differing degrees of rebound effect. Labor time saved at work due to the increased labour productivity is likely to be spent on further labor time at higher productive rates. For leisure time saving, this may simply encourage people to diversify their leisure interests to fill their generally fixed period of leisure time.

Suggested Solutions

In order to ensure that efficiency enhancing technological improvements actually reduce fuel use, the ecological economists Mathis Wackernagel and William Rees have suggested that any cost savings from efficiency gains be "taxed away or otherwise removed from further economic circulation. Preferably they should be captured for reinvestment in natural capital rehabilitation." This can be achieved through, for example, the imposition of a green tax, a cap and trade program, higher fuel taxes or the proposed "restore" approach. Policies can also directly address projected yearly consumption of energy rather than device efficiency, especially for systems where the use can be accurately projected, such as street lighting.

Environmental Full-cost Accounting

Environmental full-cost accounting (EFCA) is a method of cost accounting that traces direct costs and allocates indirect costs by collecting and presenting information about the possible environmental, social and economical costs and benefits or advantages – in short, about the "triple bottom line" – for each proposed alternative. It is also known as true-cost accounting (TCA), but, as definitions for "true" and "full" are inherently subjective, experts consider both terms problematical.

Since costs and advantages are usually considered in terms of environmental, economic and social impacts, full or true cost efforts are collectively called the "triple bottom line". A large number of standards now exist in this area including Ecological Footprint, eco-labels, and the United Nations International Council for Local Environmental Initiatives approach to triple bottom line using the ecoBudget metric. The International Organization for Standardization (ISO) has several accredited standards useful in FCA or TCA including for greenhouse gases, the ISO 26000 series for corporate social responsibility coming in 2010, and the ISO 19011 standard for audits including all these.

Because of this evolution of terminology in the public sector use especially, the term full-cost accounting is now more commonly used in management accounting, e.g. infrastructure management and finance. Use of the terms FCA or TCA usually indicate relatively conservative extensions of current management practices, and incremental improvements to GAAP to deal with waste output or resource input.

These have the advantage of avoiding the more contentious questions of social cost.

Concepts

Full-cost accounting embodies several key concepts that distinguish it from standard accounting techniques. The following list highlights the basic tenets of FCA.

Accounting for:

1. Costs rather than outlays;
2. Hidden costs and externalities;
3. Overhead and indirect costs;
4. Past and future outlays;
5. Costs according to lifecycle of the product.

Costs Rather than Outlays

Expenditure of cash to acquire or use a resource. A cost is the cash value of the resource as it is used. For example, an outlay is made when a vehicle is purchased, but the cost of the vehicle is incurred over its active life (e.g., 10 years). The cost of the vehicle must be allocated over a period of time because every year of its use contributes to the depreciation of the vehicle's value.

Hidden Costs

The value of goods and services is reflected as a cost even if no cash outlay is involved.

One community might receive a grant from a state, for example, to purchase equipment. This equipment has value, even though the community did not pay for it in cash. The equipment, therefore, should be valued in an FCA analysis.

Government subsidies in the energy and food production industries keep true costs low through artificially cheap product pricing. This price manipulation encourages unsustainable practices and further hides negative externalities endemic to fossil fuel production and modern mechanized agriculture.

Overhead and Indirect Costs

FCA accounts for all overhead and indirect costs, including those that are shared with other public agencies. Overhead and indirect costs might include legal services, administrative support, data processing, billing, and purchasing. Environmental costs as indirect costs include the full range of costs throughout the life-cycle of a product (Life cycle assessment), some of which even do not show up in the firm's bottom line. It also contains fixed overhead, fixed administration expense etc.

Past and Future Outlays

Past and future cash outlays often do not appear on annual budgets under cash accounting systems. Past (or upfront) costs are initial investments necessary to implement services such as the acquisition of vehicles, equipment, or facilities. Future (or back-end) outlays are costs incurred to complete operations such as facility closure and postclosure care, equipment retirement, and post-employment health and retirement benefits.

Examples of Full-cost Accounting

Waste Management

For example, the State of Florida uses the term full-cost accounting for its solid waste management. In this instance, FCA is a systematic approach for identifying, summing, and reporting the actual costs of solid waste management. It takes into account past and future outlays, overhead (oversight and support services) costs, and operating costs.

Integrated solid waste management systems consist of a variety of municipal solid waste (MSW) activities and paths. Activities are the building blocks of the system, which may include waste collection, operation of transfer stations, transport to waste management facilities, waste processing and disposal, and sale of byproducts. Paths are the directions that MSW follows in the course of integrated solid waste management (i.e., the point of generation through processing and ultimate disposition) and include recycling, composting, waste-to-energy, and landfill disposal. The cost of some activities is shared between paths. Understanding the costs of MSW activities is often necessary for compiling the costs of the entire solid waste system, and helps municipalities evaluate whether to provide a service itself or contract out for it. However, in considering changes that affect how much MSW ends up being recycled, composted, converted to energy, or landfilled, the analyst should focus the costs of the different paths. Understanding the full costs of each MSW path is an essential first step in discussing whether to shift the flows of MSW one way another.

Benefits

Identify the costs of MSW management

> When municipalities handle MSW services through general tax funds, the costs of MSW management can get lost among other expenditures. With FCA, managers can have more control over MSW costs because they know what the costs are.

See through the peaks and valleys in MSW cash expenditures

> Using techniques such as depreciation and amortization, FCA produces a more accurate picture of the costs of MSW programs, without the distortions that can result from focusing solely on a given year's cash expenditures.

Explain MSW costs to citizens more clearly

> FCA helps you collect and compile the information needed to explain to citizens what solid waste management actually costs. Although some people might think that solid waste management is free (because they are not billed specifically for MSW services), others might overestimate its cost. FCA can result in "bottom line" numbers that speak directly to residents. In addition, public officials can use FCA results to respond to specific public concerns.

Adopt a business-like approach to MSW management

> By focusing attention on costs, FCA fosters a more businesslike approach to MSW management. Consumers of goods and services increasingly expect value, which means an appropriate balance between quality and cost of service. FCA can help identify opportunities for streamlining services, eliminating inefficiencies, and facilitating cost-saving efforts through informed planning and decision-making.

Develop a stronger position in negotiating with vendors

> When considering privatization of MSW services, solid waste managers can use FCA to learn what it costs (or would cost) to do the work. As a result, FCA better positions public agencies for negotiations and decision-making. FCA also can help communities with publicly run operations determine whether their costs are competitive with the private sector.

Evaluate the appropriate mix of MSW services

> FCA gives managers the ability to evaluate the cost of each element of their solid waste system, such as recycling, composting, waste-to-energy, and landfilling. FCA can help managers avoid common mistakes in thinking about solid waste management, notably the error of treating avoided costs as revenues.

Fine-tune MSW programs

> As more communities use FCA and report the results, managers might be able to "benchmark" their operations to similar communities or norms. This comparison can suggest options for "re-engineering" current operations. Furthermore, when cities, counties, and towns know what it costs to manage MSW independently, they can better identify any savings that might come from working together.

Motives for Adoption

Various motives for adoption of FCA/TCA have been identified. The most significant of which tend to involve anticipating market or regulatory problems associated with ignoring the comprehensive outcome of the whole process or event accounted for. *In green economics, this is the major concern and basis for critiques of such measures as GDP.* The public sector has tended to move more towards longer term measures to avoid accusations of political favoritism towards specific solutions that seem to make financial or economic sense in the short term, but not longer term.

Corporate decision makers sometimes call on FCA/TCA measures to decide whether to initiate recalls, practice voluntary product stewardship (a form of recall at the end of a product's useful life). This can be motivated as a hedge against future liabilities arising from those who are negatively affected by the waste a product becomes. Advanced theories of FCA, such as Natural Step, focus firmly on these. According to Ray Anderson, who instituted a form of FCA/TCA at Interface Carpet, used it to rule out decisions that increase Ecological Footprint and focus the company more clearly on a sustainable marketing strategy.

The urban ecology and industrial ecology approaches inherently advocate FCA — treating the built environment as a sort of ecosystem to minimize its own wastes.

References

- Westhues, Anne; Jean Lafrance; Glen Schmidt (2001). "A SWOT analysis of social work education in Canada". Social Work Education: The International Journal. 20 (1): 35–56
- Menon, A.; et al. (1999). "Antecedents and Consequences of Marketing Strategy Making". Journal of Marketing. American Marketing Association. 63 (2): 18–40. JSTOR 1251943. doi:10.2307/1251943

- Quincy, Ronald. "SWOT Analysis: Raising capacity of your organization". Rutgers School of Social Work. Retrieved 2013-02-25

- Mata, T M (2001). "Life cycle assessment of different reuse percentages for glass beer bottles". International Journal of Life Cycle Assessment. 6 (5): 58–63. doi:10.1007/BF02978793. Retrieved 28 June 2014

- Hill, T. & R. Westbrook (1997). "SWOT Analysis: It's Time for a Product Recall". Long Range Planning. 30 (1): 46–52. doi:10.1016/S0024-6301(96)00095-7

- Spitzly, David (1997), Life Cycle Design of Milk and Juice Packaging (PDF), U.S. Environmental Protection Agency, retrieved 29 June 2014

- Roper, William E. (2006). "Strategies for building material reuse and recycle". International Journal of Environmental Technology and Management. 6 (3/4): 313–345. doi:10.1504/IJETM.2006.009000

- Cole, C; et al. (2014). "Towards a Zero Waste Strategy for an English Local Authority". Resources, Conservation and Recycling. 89: 64–75. doi:10.1016/j.resconrec.2014.05.005

- Armstrong. M. A handbook of Human Resource Management Practice (10th edition) 2006, Kogan Page , London ISBN 0-7494-4631-5

- Vogtländer J.: " Communicating the eco-efficiency of products and services by means of the eco-costs/value model", Journal of Cleaner Production 10, 2002, pp 57-67

- Davoudi, S; Evans (2005). "The Challenge of governance in regional waste planning". Environment and Planning C: Government and Policy. 23: 493–517. doi:10.1068/c42m

- Benn, Hilary; Milliband, Ed. "Guidance on how to measure and report your greenhouse gas emissions" (PDF). GOV.UK. Department for Environment, Food and Rural Affairs (UK). Retrieved 9 November 2016

- van den Bergh, Jeroen C.J.M; Grazi, Fabio (2014). "Ecological Footprint Policy? Land Use as an Environmental Indicator". Journal of Industrial Ecology. 18 (1): 10–19. ISSN 1088-1980. doi:10.1111/jiec.12045

- Frosch, R; Gallopoulos (1989). "Strategies for manufacturing". Scientific American. 261 (3): 144–152. doi:10.1038/scientificamerican0989-144

- Blake, Martin; Wijetilaka, Shehan (26 February 2015). "5 tips to grow your start-up using SWOT analysis". Sydney. Retrieved 10 August 2015

Integrating Industrial Ecology and Economy

Dematerialization is the cutting down of the quantity of materials that would be used by society. Industrial organization, circular economy, eco-efficiency, etc. are some important topics related to the subject of industrial ecology and economy. This chapter helps the readers in developing a better idea about industrial ecology and economy.

Dematerialization (Economics)

In economics, dematerialization refers to the absolute or relative reduction in the quantity of materials required to serve economic functions in society. In common terms, dematerialization means doing more with less. This concept is similar to ephemeralization as proposed by Buckminster Fuller.

In 1972, the Club of Rome in its report The Limits to Growth predicted a steadily increasing demand for material as both economies and populations grew. The report predicted that continually increasing resource demand would eventually lead to an abrupt economic collapse. Studies on material use and economic growth show instead that society is gaining the same economic growth with much less physical material required. Al Gore similarly noted in 1999 that since 1949, while the economy tripled, the weight of goods produced did not change.

By most measures, quality of life improved from 1977 to 2001. While consumer demand is constantly increasing, consumers demand services such as communication, heating and housing, and not the raw materials needed to provide these. As a result, there is incentives to provide these with less materials. Copper wire has been replaced with fiber-optics, vinyl records with MP3 players while cars, refrigerators and numerous other items have gotten lighter.

Industrial Organization

In economics, industrial organization or Industrial economy is a field that builds on the theory of the firm by examining the structure of (and, therefore, the boundaries between) firms and markets. Industrial organization adds real-world complications to the perfectly competitive model, complications such as transaction costs, limited information, and barriers to entry of new firms that may be associated with imperfect competition. It analyzes determinants of firm and market organization and behavior as between competition and monopoly, including from government actions.

There are different approaches to the subject. One approach is descriptive in providing an overview of industrial organization, such as measures of competition and the size-concentration of firms in an industry. A second approach uses microeconomic models to explain internal firm organization and market strategy, which includes internal research and development along with issues of internal reorganization and renewal. A third aspect is oriented to public policy as to economic regulation, antitrust law, and, more generally, the economic governance of law in defining property rights, enforcing contracts, and providing organizational infrastructure.

The subject has a theoretical side and a practical side. According to one textbook: "On one plane the field is abstract, a set of analytical concepts about competition and monopoly. On a second plane the topic is about real markets, teeming with the excitement and drama of struggles among real firms" (Shepherd, W.; 1985; 1).

The extensive use of game theory in industrial economics has led to the export of this tool to other branches of microeconomics, such as behavioral economics and corporate finance. Industrial organization has also had significant practical impacts on antitrust law and competition policy.

The development of industrial organization as a separate field owes much to Edward Chamberlin, Edward S. Mason, J. M. Clark, and particularly Joe S. Bain among others.

Assessments of the subject have differed over time. The preface to a related research volume in 1972 remarked on *Whither industrial organization?*: "That all is not well with this in this once flourishing field is readily apparent." A response came 15 years later: "[T]oday's verdict is that industrial organization is alive and well and the queen of applied microeconomics."

Subareas

The Journal of Economic Literature (JEL) classification codes are one way of representing the range of economics subjects and subareas. There, Industrial Organization, one of 20 primary categories, has 9 secondary categories, each with multiple tertiary categories. The secondary categories are listed below with corresponding available article-preview links of The New Palgrave Dictionary of Economics Online and footnotes to their respective JEL-tertiary categories and associated New-Palgrave links.

JEL: L1 – Market Structure, Firm Strategy, and Market Performance

JEL: L2 – Firm Objectives, Organization, and Behavior

JEL: L3 – Non-profit organizations and Public enterprise

JEL: L4 – Antitrust Issues and Policies

JEL: L5 – Regulation and Industrial policy

JEL: L6 – Industry Studies: Manufacturing

JEL: L7 – Industry Studies: Primary Products and Construction

JEL: L8 – Industry Studies: Services

JEL: L9 – Industry Studies: Transportation and Utilities

Market Structures

The common market structures studied in this field are the following:

- Perfect competition
- Monopolistic competition
- Duopoly
- Oligopoly
- Oligopsony
- Monopoly
- Monopsony

Areas of Study

Industrial organization investigates the outcomes of these market structures in environments with

- Price discrimination
- Product differentiation
- Durable goods
- Experience goods
- Secondary markets, which can affect the behaviour of firms in primary markets.

- Collusion
- Signalling, such as warranties and advertising.
- Mergers and acquisitions
- Entry and Exit

History of the Field

A 2009 book *Pioneers of Industrial Organization* traces the development of the field from Adam Smith to recent times and includes dozens of short biographies of major figures in Europe and North America who contributed to the growth and development of the discipline.

Circular Economy

A circular economy is a regenerative system in which resource input and waste, emission, and energy leakage are minimised by slowing, closing, and narrowing material and energy loops. This can be achieved through long-lasting design, maintenance, repair, reuse, remanufacturing, refurbishing, and recycling. This is contrast to a linear economy which is a 'take, make, dispose' model of production.

Scope

The term encompasses more than the production and consumption of goods and services, including a shift from fossil fuels to the use of renewable energy, and the role of diversity as a characteristic of resilient and productive systems. It includes discussion of the role of money and finance as part of the wider debate, and some of its pioneers have called for a revamp of economic performance measurement tools.

"The concept of a circular economy (CE) has been first raised by two British environmental economists David W. Pearce and R. Kerry Turner in 1989. In *Economics of Natural Resources and the Environment*, they pointed out that a traditional open-ended economy was developed with no built-in tendency to recycle, which was reflected by treating the environment as a waste reservoir". The circular economy is grounded in the study of feedback-rich (non-linear) systems, particularly living systems. A major outcome of this is the notion of optimising systems rather than components, or the notion of 'design for fit'. As a generic notion it draws from a number of more specific approaches including cradle to cradle, biomimicry, industrial ecology, and the 'blue economy'.

Moving Away from the Linear Model

Linear "take, make, dispose" industrial processes and the lifestyles that feed on them deplete finite reserves to create products that end up in landfills or in incinerators.

This realisation triggered the thought process of a few scientists and thinkers, including Walter R. Stahel, an architect, economist, and a founding father of industrial sustainability. Credited with having coined the expression "Cradle to Cradle" (in contrast with "Cradle to Grave", illustrating our "Resource to Waste" way of functioning), in the late 1970s, Stahel worked on developing a "closed loop" approach to production processes, co-founding the Product-Life Institute in Geneva more than 25 years ago. In the UK, Steve D. Parker researched waste as a resource in the UK agricultural sector in 1982, developing novel closed loop production systems mimicking, and integrated with, the symbiotic biological ecosystems they exploited.

Emergence of the Idea

In their 1976 Hannah Reekman research report to the European Commission, "The Potential for Substituting Manpower for Energy", Walter Stahel and Genevieve Reday sketched the vision of an economy in loops (or circular economy) and its impact on job creation, economic competitiveness, resource savings, and waste prevention. The report was published in 1982 as the book *Jobs for Tomorrow: The Potential for Substituting Manpower for Energy*.

Considered as one of the first pragmatic and credible sustainability think tanks, the main goals of Stahel's institute are product-life extension, long-life goods, reconditioning activities, and waste prevention. It also insists on the importance of selling services rather than products, an idea referred to as the "functional service economy" and sometimes put under the wider notion of "performance economy" which also advocates "more localisation of economic activity".

In broader terms, the circular approach is a framework that takes insights from living systems. It considers that our systems should work like organisms, processing nutrients that can be fed back into the cycle—whether biological or technical—hence the "closed loop" or "regenerative" terms usually associated with it.

The generic Circular Economy label can be applied to, and claimed by, several different schools of thought, that all gravitate around the same basic principles which they have refined in different ways. The idea itself, which is centred on taking insights from living systems, is hardly a new one and hence cannot be traced back to one precise date or author, yet its practical applications to modern economic systems and industrial processes have gained momentum since the late 1970s, giving birth to four prominent movements, detailed below. The idea of circular material flows as a model for the economy was presented in 1966 by Kenneth E. Boulding in his paper, The Economics of the Coming Spaceship Earth. Promoting a circular economy was identified as national policy in China's 11th five-year plan starting in 2006. The Ellen MacArthur Foundation, an independent charity established in 2010, has more recently outlined the economic opportunity of a circular economy. As part of its educational mission, the Foundation has worked to bring together complementary schools of thought and create a coherent framework, thus giving the concept a wide exposure and appeal.

Most frequently described as a framework for thinking, its supporters claim it is a coherent model that has value as part of a response to the end of the era of cheap oil and materials and can contribute to the transition to a low carbon economy. In line with this, a circular economy can contribute to meet the COP 21 Paris Agreement. The emissions reduction commitments made by 195 countries at the COP 21 Paris Agreement, are not sufficient to limit global warming to 1.5 °C. To reach the 1.5 °C ambition it is estimated that additional emissions reductions of 15 billion tonnes CO_2 per year need to be achieved by 2030. Circle Economy and Ecofys estimated that circular economy strategies may deliver emissions reductions that could basically bridge the gap by half.

Sustainability

The circular economy seems intuitively to be more sustainable than the current linear economic system. The reduction of resource inputs into and waste and emission leakage out of the system reduces resource depletion and environmental pollution. However, these simple assumptions are not sufficient to deal with the involved systemic complexity and disregards potential trade-offs. For example, the social dimension of sustainability seems to be only marginally addressed in many publications on the Circular Economy, and there are cases that require different or additional strategies, like purchasing new, more energy efficient equipment. By reviewing the literature, a team of researchers from Cambridge and TU Delft could show that there are at least eight different relationship types between sustainability and the circular economy:

1. Conditional relation

2. Strong conditional relation

3. Necessary but not sufficient conditional relation

4. Beneficial relationship

5. Subset relation (structured and unstructured)

6. Degree relation

7. Cost-benefit/trade-off relation

8. Selective relation

Key Elements

With a surge in popularity, many circular principles are available, varying widely depending on the problems being addressed, the audience, or the lens through which the author views the world. There are at least the following key elements to be identified within a circular economy.

Prioritise Regenerative Resources

Ensure renewable, reusable, non-toxic resources are utilised as materials and energy in an efficient way. Ultimately the system should aim to run on 'current sunshine' and generate energy through renewable sources. An example of this principle is The Biosphere Rules framework for closed-loop production which identifies Power Autonomy as one of nature's principles for sustainable manufacturing. It requires that energy efficiency be first maximized so that renewable energy becomes economical. It also requires that materials need to be non-toxic to be able to recirculate without causing harm to the living environment.

Use Waste as a Resource

The second element aims to utilise waste streams as a source of secondary resources and recover waste for reuse and recycling and is grounded on the idea that waste does not exist. It is necessary here to design out waste, meaning that both the biological and technical components (nutrients) of a product are designed intentionally in such a way that waste streams are minimalized.

Design For the Future

Account for the systems perspective during the design process, to use the right materials, to design for appropriate lifetime and to design for extended future use. Meaning that a product is designed to fit within a materials cycle, can easily be dissembled and can easily be used with a different purpose. Hereby one could consider strategies like emotionally durable design. It should be stressed that there is not something like one ideal blueprint for future design. Modularity, versatility and adaptiveness are to be prioritised in an uncertain and fast evolving world, meaning that diverse products, materials, and systems, with many connections and scales are more resilient in the face of external shocks, than monotone systems built simply for efficiency.

Preserve and Extend what's Already Made

While resources are in-use, maintain, repair and upgrade them to maximise their lifetime and give them a second life through take back strategies when applicable. This could mean that a product is accompanied with a pre-thought maintenance programme to maximise its lifetime, including a buyback program and supporting logistics system. Second hand sales or refurbish programs also falls within this element.

Collaborate to Create Joint Value

Within a circular economy, one should work together throughout the supply chain, internally within organisations and with the public sector to increase transparency and create joint value. For the business sector this calls for collaboration within the supply chain and cross-sectoral, recognising the interdependence between the different market players. Governments can support this by creating the right incentives, for example via common standards within a regulatory framework and provide business support.

Incorporate Digital Technology

Track and optimise resource use and strengthen connections between supply chain actors through digital, online platforms and technologies that provide insights. It also encompasses virtualized value creation and delivering, for example via 3D printers, and communicating with customers virtually.

Prices or other Feedback Mechanisms should Reflect Real Costs

In a circular economy, prices act as messages, and therefore need to reflect full costs in order to be effective. The full costs of negative externalities are revealed and taken into account, and perverse subsidies are removed. A lack of transparency on externalities acts as a barrier to the transition to a circular economy.

The Circular Economy Framework

The circular economy is a framework that draws upon and encompasses principles from:

Systems Thinking

The ability to understand how things influence one another within a whole. Elements are considered as 'fitting in' their infrastructure, environment and social context. Whilst a machine is also a system, systems thinking usually refers to nonlinear systems: systems where through feedback and imprecise starting conditions the outcome is not necessarily proportional to the input and where evolution of the system is possible: the system can display emergent properties. Examples of these systems are all living systems and any open system such as meteorological systems or ocean currents, even the orbits of the planets have nonlinear characteristics.

Understanding a system is crucial when trying to decide and plan (corrections) in a system. Missing or misinterpreting the trends, flows, functions of, and human influences on, our socio-ecological systems can result in disastrous results. In order to prevent errors in planning or design an understanding of the system should be applied to the whole and to the details of the plan or design. The Natural Step created a set of systems conditions (or sustainability principles) that can be applied when designing for (parts of) a circular economy to ensure alignment with functions of the socio-ecological system.

The concept of the circular economy has previously been expressed as the circulation of money versus goods, services, access rights, valuable documents, etc., in macroeconomics. This situation has been illustrated in many diagrams for money and goods circulation associated with social systems. As a system, various agencies or entities are connected by paths through which the various goods etc., pass in exchange for money. However, this situation is different from the circular economy described above, where the flow is unilinear - in only one direction, that is, until the recycled goods again are spread over the world.

Biomimicry

Janine Benyus, author of "Biomimicry: Innovation Inspired by Nature", defines her approach as "a new discipline that studies nature's best ideas and then imitates these designs and processes to solve human problems. Studying a leaf to invent a better solar cell is an example. I think of it as "innovation inspired by nature. Biomimicry relies on three key principles:

- Nature as model: Biomimicry studies nature's models and emulates these forms, processes, systems, and strategies to solve human problems.

- Nature as measure: Biomimicry uses an ecological standard to judge the sustainability of our innovations.

- Nature as mentor: Biomimicry is a way of viewing and valuing nature. It introduces an era based not on what we can extract from the natural world, but what we can learn from it.

Industrial Ecology

Industrial Ecology is the study of material and energy flows through industrial systems. Focusing on connections between operators within the "industrial ecosystem", this approach aims at creating closed loop processes in which waste is seen as input, thus eliminating the notion of undesirable by-product. Industrial ecology adopts a systemic - or holistic - point of view, designing production processes according to local ecological constraints whilst looking at their global impact from the outset, and attempting to shape them so they perform as close to living systems as possible. This framework is sometimes referred to as the "science of sustainability", given its interdisciplinary nature, and its principles can also be applied in the services sector. With an emphasis on natural capital restoration, Industrial Ecology also focuses on social wellbeing.

Cradle to Cradle

Created by Walter R. Stahel, a Swiss architect, who graduated from the Swiss Federal Institute of Technology Zürich in 1971. He has been influential in developing the field of sustainability, by advocating 'service-life extension of goods - reuse, repair, remanufacture, upgrade technologically' philosophies as they apply to industrialised economies. He co-founded the Product Life Institute in Geneva, Switzerland, a consultancy devoted to developing sustainable strategies and policies, after receiving recognition for his prize winning paper 'The Product Life Factor' in 1982. His ideas and those of similar theorists led to what is now known as the circular economy in which industry adopts the reuse and service-life extension of goods as a strategy of waste prevention, regional job creation and resource efficiency in order to decouple wealth from resource consumption, that is to dematerialise the industrial economy.

Cooper (2005) proposed a theoretical model to illustrate the significance of product life span in a progress towards sustainable consumption. The longer product life spans could contribute to eco-efficiency and sufficiency, thus, slowing the consumption in order to progress towards sustainable consumption.

Blue Economy

Initiated by former Ecover CEO and Belgian entrepreneur Gunter Pauli, derived from the study of natural biological production processes the official manifesto states, "using the resources availablethe waste of one product becomes the input to create a new cash flow". Based on 21 founding principles, the Blue Economy insists on solutions being determined by their local environment and physical / ecological characteristics, putting the emphasis on gravity as the primary source of energy - a point that differentiates this school of thought from the others within the Circular Economy. The report - which doubles as the movement's manifesto - describes "100 innovations which can create 100 million jobs within the next 10 years", and provides many example of winning South-South collaborative projects, another original feature of this approach intent on promoting its hands-on focus.

Biosphere Rules

The Biosphere Rules is a framework for implementing closed loop production processes. They derived from nature systems and translated for industrial production systems. The five principles are Materials Parsimony, Value Cycling, Power Autonomy, Sustainable Product Platforms and Function Over Form.

Towards the Circular Economy

In January 2012, a report was released entitled *Towards the Circular Economy: Economic and business rationale for an accelerated transition*. The report, commissioned by the Ellen MacArthur Foundation and developed by McKinsey & Company, was the first of its kind to consider the economic and business opportunity for the transition to a restorative, circular model. Using product case studies and economy-wide analysis, the report details the potential for significant benefits across the EU. It argues that a subset of the EU manufacturing sector could realise net materials cost savings worth up to $630 billion annually towards 2025—stimulating economic activity in the areas of product development, remanufacturing and refurbishment. *Towards the Circular Economy* also identified the key building blocks in making the transition to a circular economy, namely in skills in circular design and production, new business models, skills in building cascades and reverse cycles, and cross-cycle/cross-sector collaboration.

In January 2015 a *Definitive Guide to The Circular Economy* was published by Coara with the specific aim to raise awareness amongst the general population of the environmental problems already being caused by our "throwaway culture". Waste Electrical and Electronic Equipment (WEEE,) in particular, is contributing to excessive use of landfill sites across the globe in which society is both discarding valuable metals but also dumping toxic compounds that are polluting the surrounding land and water supplies. Mobile devices and computer hard drives typically contain valuable metals such as silver and copper but also hazardous chemicals such as lead, mercury and cadmium. Consumers are unaware of the environmental significance of upgrading their mobile phones, for instance, on such a frequent basis but could do much to encourage manufacturers to start to move away from the wasteful, polluting linear economy towards are sustainable circular economy.

Impact in Europe

On 17 December 2012, the European Commission published a document entitled *Manifesto for a Resource Efficient Europe*. This manifesto clearly stated that "In a world with growing pressures on resources and the environment, the EU has no choice but to go for the transition to a resource-efficient and ultimately regenerative circular economy." Furthermore, the document highlighted the importance of "a systemic change in the use and recovery of resources in the economy" in ensuring future jobs and competitiveness, and outlined potential pathways to a circular economy, in innovation and investment, regulation, tackling harmful subsidies, increasing opportunities for new business models, and setting clear targets.

The European environmental research and innovation policy aims at supporting the transition to a circular economy in Europe, defining and driving the implementation of a transformative agenda to green the economy and the society as a whole, to achieve a truly sustainable development. Research and innovation in Europe are financially supported by the programme Horizon 2020, which is also open to participation worldwide.

The European Commission introduced a Circular Economy proposal in 2015. Historically, the policy debate in Brussels mainly focused on waste management which is the second half of the cycle, and very little is said about the first half: eco-design. To draw the attention of policymakers and other stakeholders to this loophole, the Ecothis. An EU campaign was launched raising awareness about the economic and environmental consequences of not including eco-design as part of the circular economy package.

Circular Business Model

A circular economy calls upon opportunities to create greater value and align incentives through business models that build on the interaction between products and services. Linder and Williander describe a circular business model as "a business model in which the conceptual logic for value creation is based on utilizing the economic value retained in products after use in the production of new offerings".

Basically this means that a circular business model is not focused merely on selling a product, but encompasses a shift in thinking about value proposition, bringing forward a whole range of different business models to be used. To mention just a few examples: product-service systems, virtualized services, and collaborative consumption which encompasses the sharing economy. This comprises both the incentives and benefits offered to customers for bringing back used products and a change in revenue streams, comprising payments for a circular product or service, or payments for delivered availability, usage, or performance related to the product-based service offered.

These new ways of doing business require businesses to create an attractive business model for financiers, and financiers to change the way they perceive the risks and opportunities associated with these models. To help businesses position themselves in a circular context and develop future strategies for doing business in a circular economy, the Value Hill has been created. The Value Hill proposes a categorisation based on the lifecycle phases of a product: pre-, in- and post- use. This allows businesses to position themselves on the Value Hill and understand possible circular strategies they can implement as well as identify missing partners in their circular network. The Value Hill provides an overview of the circular partners and collaborations essential to the success of a circular value network.

Mateusz Lewandowski provides a proposition to address this need to design circular business models and presents a an extension of the framework from Osterwalder and Pigneur, namely the circular business model canvas (CBMC). The CBMC consists of eleven building blocks, encompassing not only traditional components with minor modifications, but also material loops and adaptation factors. Those building blocks allow the designing of a business model according to the principles of circular economy.

Eco-efficiency

Over the years, as countries and regions around the world began to develop, it slowly became evident that industrialization and economic growth come hand in hand with environmental degradation. Eco-Efficiency has been proposed as one of the main tools to promote a transformation from unsustainable development to one of sustainable development. It is based on the concept of creating more goods and services while using fewer resources and creating less waste and pollution. "It is measured as the ratio between the (added) value of what has been produced (e.g. GDP) and the (added) environment impacts of the product or service (e.g. S02 emissions)." The term was coined by the World Business Council for Sustainable Development (WBCSD) in its 1992 publication "Changing Course," and at the 1992 Earth Summit, eco-efficiency was endorsed as a new business concept and means for companies to implement Agenda 21 in the private sector. Ergo the term has become synonymous with a management philosophy geared towards sustainability, combing ecological and economic efficiency.

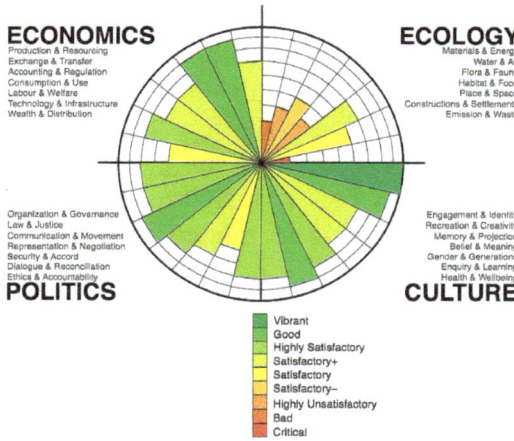

CIRCLES OF SUSTAINABILITY

History

Although eco-efficiency is a rather new method, the idea is not. In the early 1970s Paul R. Ehrlich and John Holdren developed the lettering formula I = PAT to describe the impact of human activity on the environment. Furthermore, the concept of eco-efficiency was first described by McIntyre and Thornton in 1978, but it wasn't until 1992, when the term was formally coined and widely publicized by Stephan Schmidheiny in "Changing Course". Schmidheiny set out "to change the perception of industry as being part of the problem of environmental degradation to the reality of its becoming part—a key part—of the solution for sustainability and global development." The major drivers in the early phase of eco-efficiency's development were the "forward-looking managers and thinkers in 3M and Dow." It was their involvement which catapulted eco-efficiency from a brilliant idea to a workable concept. The results of the WBCSD's work creating the "linkage between environmental performance and the bottom line was published in 1997 in its report Environmental Performance and Shareholder Value."

Methods

According to the WBCSD definition, eco-efficiency is achieved through the delivery of "competitively priced goods and services that satisfy human needs and bring quality of life while progressively reducing environmental impacts of goods and resource intensity throughout the entire life-cycle to a level at least in line with the Earth's estimated carrying capacity." It works by implementing 4 main types of ratios.

> "The first two are environmental productivity and its inverse, environmental intensity of production, referring to the realm of production. The second pair, environmental improvement cost and its inverse, environmental cost-effectiveness, are defined from an environmental improvements measures point-of-view."

The ratios may be applied to any unit comprising economic activities because such activities always relate to cost and value, "and having some physical substrate, always influence the environment." Furthermore, there are two different levels upon which to orchestrate the ratios: "micro" and "macro". There are three different methods to determine eco-efficiency at the micro-level. First, "incremental eco-efficiency", which "specifies the effects of the total value of a product system or sector and its total concomitant environmental effects." Second, an analysis method nicknamed "win-win", which "gives a comparison between a historical reference situation and potentially new situations based on the use of new technologies." It should be noted that the win-win micro-method is limited because it cannot give a concrete answer on the question of whether it improves overall environmental performance. And the third is "difference eco-efficiency", which is similar to the win-win variant, but removes all irrelevant alternatives to heighten potential for optimal technologies while comparing two alternatives. Now the macro-level is much less defined and has shown less accurate results. However, "the ultimate aim of eco-efficiency analysis is to help move micro-level decision making into macro-level optimality." The main goal in years to come is to create headline indicators to carry out macro-level analysis at a country/world scale.

There are two life-cycle assessment (LCA) -based calculation systems on eco-efficiency: the Analysis Method of BASF, and the method of the Eco-costs value ratio of the Delft University of Technology.

Uses

The reduction in ecological impacts translates into an increase in resource productivity, which in turn can create a competitive advantage. According to the WBCSD, critical aspects of eco-efficiency are:

- A reduction in the material intensity of goods or services;
- A reduction in the energy intensity of goods or services;
- Reduced dispersion of toxic materials;
- Improved recyclability;
- Maximum use of renewable resources;
- Greater durability of products;
- Increased service intensity of goods and services.

Strategies that have been linked to eco-efficiency include "Factor 4" and "Factor 10", which call for specific reductions in resource use, "natural capitalism", which incorporates eco-efficiency as part of a broader strategy, and the "cradle-to-cradle" movement, which claims to go beyond eco-efficiency in abolishing the very idea of waste. According to Boulanger, all versions of eco-efficiency share four key characteristics:

- Confidence in technological innovation as the main solution to un-sustainability;
- Reliance on business as the principal actor of transformation. The emphasis is on firms designing new products, shifting to new production processes, and investing in R&D, etc., more than on the retailer or the consumer, let alone the citizen.
- Trust in markets (if they are functioning well);
- "Growthphilia": there is nothing wrong with growth as such.

The view that improvements in eco-efficiency are sufficient for achieving sustainability has been challenged by Huesemann and Huesemann, who demonstrate using extensive historical evidence that increases in technological efficiency have not reduced overall resource use and pollution. Moreover, with "cradle-to-cradle", growth is conducive to sustainability per se. This broader concept is called "Sustainable Production and Consumption" (SPC). "This concept involves changes in production and consumption patterns that lead to sustainable use of natural resources;" business has taken a key role in accelerating the use of this concept because businesses both consume and produce. Eco-efficiency is routinely a concept used because it combines performance along two of the three axes of sustainable development, making it easier for academics and leading thinkers to tease out the associated social issues.

Examples

Furthermore, eco-efficiency is also a very useful tool because it can adapt and flex to be fit different sizes of companies, while also maintaining relevance with the larger scale of government and national policies. For example, larger national players such as the Organisation for Economic Co-operation and Development (OECD 2002), European Commission (EU 2005), European Environment Agency (EEA) and the National Round Table on the Environment and the Economy (NRTEE) have all recognized that eco-efficiency is a practical approach that businesses should adopt in setting and achieving their environmental performance objectives. It has be proven to heighten market values for firms, serve as an effective management tool for governments, benefit civil society, and increase quality of life. "It does this by changing industrial processes, creating new products and changing and influencing markets with new ideas and with new rules." More people aim to get more value for their money in the market, while also enjoying a better environment.

Recently, there has also been use of eco-efficiency in more non-traditional ways, such as a use in banks to integrate environmental criteria into their credit-approval process; looking at "eco-integrated economic risks of a customer." And is also being implemented as marketing advantages where, "eco-efficient choices are always preferred," especially in service sectors such as tourism.

Ecoleasing

Ecoleasing is a system in which goods (mainly from the technical cycle, i.e. appliances,) are rented to a client for a certain period of time after which he returns the goods so the company that made it can recycle the materials.

Terminology

The term ecoleasing has been used by William McDonough and Michael Braungart in their book Cradle_to_Cradle: Remaking the Way We Make Things. It is used to distinct itself from regular leasing in that:

- the operation is similar to regular purchasing of goods, so not requiring a contract to be made up as with leasing
- it is done with appliances and other products used for the household, rather than with land or very expensive products (for example: cars)
- the period of time the product is rented would be about the same as the lifespan of the product, so it can only be rented once before it is taken back by the company to recover the materials (and to create another product with it)

Examples

Ecoleasing can for instance be done with TV's. Instead of the consumer purchasing the

TV, by ecoleasing it, he is entitled to say 10 000 service hours. After this, he can send the TV back to the company.

Advantages

- Since materials are reclaimed, fewer or no materials end up in landfills, or require other forms of waste disposal. As such, it is quite environmental.
- Materials are recovered by which the company can make new products, so the material costs for this new product are much lower for the company.
- Since new products can be made at a lower expense, the sale price of these products can also be comparatively lower than similar products made by the competition (if they use a system of purchasing the goods)

Eco-investing

Eco-investing or green investing, is a form of socially responsible investing where investments are made in companies that support or provide environmentally friendly products and practices. These companies encourage (and often profit from) new technologies that support the transition from carbon dependence to more sustainable alternatives.

As industries' environmental impacts become more apparent, green topics have not only taken center stage in pop culture, but the financial world as well. In the 1990s many investors "began to look for those companies that were better than their competitors in terms of managing their environmental impact." While some investors still focus their funds to avoid only "the most egregious polluters," the emphasis for many investors has switched to changing "the way money is used," and using "it in a positive, transformative way to get us from where we are now ultimately to a truly sustainable society."

The Global Climate Prosperity Scoreboard – launched by Ethical Markets Media and The Climate Prosperity Alliance to monitor private investments in green companies – estimated that over $1.248 trillion has been invested in solar, wind, geothermal, ocean/hydro and other green sectors since 2007. This number represents investments from North America, China, India, and Brazil, as well at other developing countries.

Eco/Green Investing Versus Socially Responsible Investing

While many eco-investments may be considered socially responsible investments, and vice versa, the two are not mutually inclusive. Socially responsible investing is the practice of investing only in those companies which satisfy a certain moral or ethical criteria. This may include companies with an interest in the environment, but also supports various other social and religious issues.

Eco-investing narrows in on the interests of sustainable environmental issues. Specifically, eco-investments focus on companies who work on renewable energy and clean technologies.

Eco-investing Sectors

There are several sectors that fall under the eco-investing umbrella. Renewable energy refers to both solar, wind, tidal current, wave and conventional hydro technology. This includes companies that build solar panels or wind turbines, or the raw materials and services that contribute to these technologies It also refers to Energy Storage – companies that develop and use technologies to store large amounts of energy, particularly renewable energies. A good example of this is the fuel cells used in hybrid cars. Also under the renewable energy sector are Biofuels. This group includes companies that use or supply biological resources (like algae, corn or waster wood) to create energy or fuel. Other technologies that are included in the renewable energy group are: Geothermal (companies who use or convert heat to electric energy) and Hydroelectricity (companies who harness water energy to make electricity).

The Buildings and Efficiency sector refers to companies that manufacture green building materials or energy-efficient services in the world of engineering and architecture. Green building materials include energy-efficient glass, insulation, and lighting among others. Recycling companies and energy conservation companies also fall under this sector.

The Eco Living sector refers to companies that offer sustainable goods and services for healthy living. This includes organic farming, green pesticides, health care and pharmaceuticals.

Green investment has significantly grown in the UK and there are now 136 funds listed on the Worldwise Investor fund library under the themes: Agriculture, Carbon, Clean Energy, Forestry, Environmental, Multi-thematic and Water. All of these funds account for around £21.8bn in the UK.

References

- Guenster, N.; Bauer, R.; Derwall, J.; Koedijk, K. (2011). "The economic value of corporate eco-efficiency". European Financial Management. 17 (4): 679–704. doi:10.1111/j.1468-036X.2009.00532.x
- Rosenberg, Nathan (1982). Inside the Black Box: Technology and Economics. Cambridge, New York: Cambridge University Press. p. 72. ISBN 0-521-27367-6
- "A review of the circular economy in China: moving from rhetoric to implementation". Journal of Cleaner Production. 42: 215–227. 2012. doi:10.1016/j.jclepro.2012.11.020. Retrieved 12 February 2016
- Zhijun, F; Nailing, Y (2007). "Putting a circular economy into practice in China" (PDF). Sustain Sci. 2: 95–101. doi:10.1007/s11625-006-0018-1

- Achterberg, Elisa; Hinfelaar, Jeroen; Bocken, Nancy. "Master Circular Business With The Value Hill". Circle Economy, Sustainable Finance Lab, Nuovalente, TU Delft. Retrieved 20 April 2017

- Cooper, Tim (2005). "Slower Consumption Reflections on Product Life Spans and the "Throwaway Society"". Journal of Industrial Ecology. 9 (1-2): 51–67. doi:10.1162/1088198054084671 – via Willey

- Linder, Marcus; Williander, Mats (2015-01-01). "Circular Business Model Innovation: Inherent Uncertainties". Business Strategy and the Environment: n/a–n/a. ISSN 1099-0836. doi:10.1002/bse.1906

- David W. Pearce and R. Kerry Turner (1989). Economics of Natural Resources and the Environment. Johns Hopkins University Press. ISBN 978-0801839870

- Blok, Kornelis; Hoogzaad, Jelmer; Ramkumar, Shyaam; Ridley, Shyaam; Srivastav, Preeti; Tan, Irina; Terlouw, Wouter; de Wit, Terlouw. "Implementing Circular Economy Globally Makes Paris Targets Achievable". CircleEconomy. Circle Economy, Ecofys. Retrieved 20 April 2017

- Ehrenfeld, J. R. (2005). "Eco-efficiency: Philosophy, theory and tools". Journal of Industrial Ecology. 9 (4): 6–8. doi:10.1162/108819805775248070

- Lewandowski, Mateusz (2016-01-18). "Designing the Business Models for Circular Economy—Towards the Conceptual Framework". Sustainability. 8 (1): 43. doi:10.3390/su8010043

- Sinkin, C.; Wright, C. J.; Burnett, R. D. (2008). "Eco-efficiency and firm value". Journal of Accounting and Public Policy. 27 (2): 167–176. doi:10.1016/j.jaccpubpol.2008.01.003

- McIntyre, R.; Thornton, J. (1978). "On the environmental efficiency of economic systems". Soviet Studies. 30 (2): 173–192. doi:10.1080/09668137808411179

Materials Flow in Industrial Ecology

The transportation of raw materials and other vital industrial elements from one place to another is known as material flow. The typical tools used in the process are AnyLogic, plant simulation for production systems and AutoMod for logistics systems. Other themes include supply chain management, industrial metabolism, etc. This chapter will provide an integrated understanding of materials flow in industrial ecology.

Material Flow

Material flow is the description of the transportation of raw materials, pre-fabricates, parts, components, integrated objects and final products as a flow of entities. The term applies mainly to advanced modeling of supply chain management.

As industrial material flow can easily become very complex, several different specialized simulation tools have been developed for complex systems. Typical tools are:

- AnyLogic
- AutoMod for logistics systems
- Plant Simulation for production system

Material Flow Management

Material flow management (MFM) is a method of efficiently managing materials.

Material flow management is the goal oriented, efficient use of materials, material streams and energy. The goals are given by ecological and economical areas and by observing social aspects. (in "Protection of human beings and environment", by an Enquete Commission of the German Bundestag)

This triple jump of environmental, social and economical orientation makes MFM a tool of high importance in the field of Sustainable Development (SD) and Circular Economy (CE). Seen historically Material Flow Management is a relatively new tool that can be understood as an implementation-orientated advancement of the methodology of Material Flow Analysis (MFA). MFM was established as a policy tool after the UN conference in Rio de Janeiro 1992. The German "Bundestag" clearly outlined the targets and specific goals of MFM in a special report by an Enquete Commission.

Material Flow Analysis

Material flow analysis (MFA) (also referred to as substance flow analysis (SFA)) is an analytical method to quantify flows and stocks of materials or substances in a well-defined system. MFA is an important tool to study the bio-physical aspects of human activity on different spatial and temporal scales. It is considered a core method of industrial ecology or anthropogenic, urban, social and industrial metabolism. MFA is used to study material, substance, or product flows across different industrial sectors or within ecosystems. MFA can also be applied to a single industrial installation, for example, for tracking nutrient flows through a waste water treatment plant. When combined with an assessment of the costs associated with material flows this business-oriented application of MFA is called Material Flow Cost Accounting. MFA is an important tool to study the circular economy and to devise material flow management. Since the 1990s, the number of publications related to material flow analysis has grown steadily. Peer-reviewed journals that publish MFA-related work include the Journal of Industrial Ecology, Ecological Economics, Environmental Science and Technology, and Resources, Conservation, and Recycling.

Methodology of MFA

Motivation

Human needs such as shelter, food, transport, or communication require materials like wood, starch, sugar, iron and steel, copper, or semiconductors. As society develops and economic activity expands, material production, use, and disposal increase to a level where unwanted impacts on environment and society cannot be neglected anymore, neither locally nor globally. Material flows are at the core of local environmental problems such as leaching from landfills or oil spills. Rising concern about global warming puts a previously unimportant waste flow, carbon dioxide, on top of the political and scientific agenda. The gradual shift from primary material production to urban mining in developed countries requires a detailed assessment of in-use and obsolete stocks of materials within human society. Scientists, industries, government bodies, and NGOs therefore need a tool that complements economic accounting and modelling. They need a systematic method to keep track of and display stocks and flows of the materials entering, staying within, and leaving the different processes in the anthroposphere. Material flow analysis is such a method.

Basic Principles

MFA is based on two fundamental and well-established scientific principles, the systems approach and mass balance. The system definition is the starting point of every MFA study.

System Definition

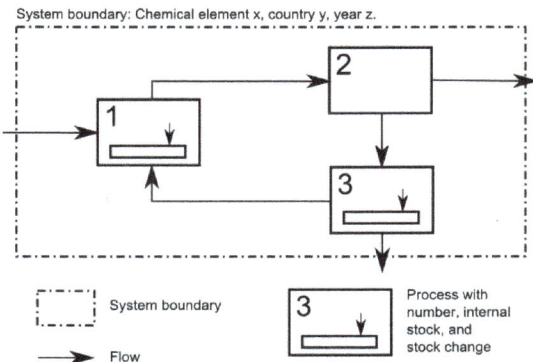

Basic MFA system without quantification.

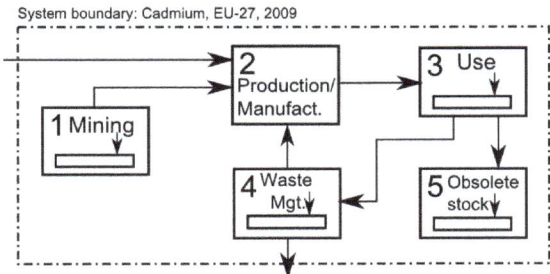

A more general MFA system without quantification.

An MFA system is a model of an industrial plant, an industrial sector or a region of concern. The level of detail of the system model is chosen to fit the purpose of the study. An MFA system always consists of the system boundary, one or more processes, material flows between processes, and stocks of materials within processes. Physical exchange between the system and its environment happens via flows that cross the system boundary. Contrary to chemical engineering, where a system represents a specific industrial installation, systems and processes in MFA can represent much larger and more abstract entities as long as they are well-defined. The explicit system definition helps the practitioner to locate the available quantitative information in the system, either as stocks within certain processes or as flows between processes. An MFA system description can be refined by disaggregating processes or simplified by aggregating processes.

Next to specifying the arrangement of processes, stocks, and flows in the system definition, the practitioner also needs to indicate the scale and the indicator element or material of the system studied. The spatial scale describes the geographic entity that is covered by the system. A system representing a certain industrial sector can be applied to the USA, China, certain world regions, or the world as a whole. The temporal scale describes the point in time or the time span for which the system is quantified. The indicator element or material of the system is the physical entity that is measured and for which the mass balance holds. As the name says, an indicator element is a certain chemical element such as cadmium or a substance such as CO_2. In general, a material or a product can also be used as indicator as long as a process balance can be established for it. Examples of more general indicators are goods such as passenger cars, materials like steel, or other physical quantities such as energy.

MFA requires practitioners to make precise use of the terms 'material', 'substance', or 'good'.

- A chemical element is "a pure chemical substance consisting of one type of atom distinguished by its atomic number".

- A substance is "any (chemical) element or compound composed of uniform units. All substances are characterised by a unique and identical constitution and are thus homogeneous."

- A good is defined as "economic entity of matter with a positive or negative economic value. Goods are made up of one or several substances".

- The term material in MFA "serves as an umbrella term for both substances and goods."

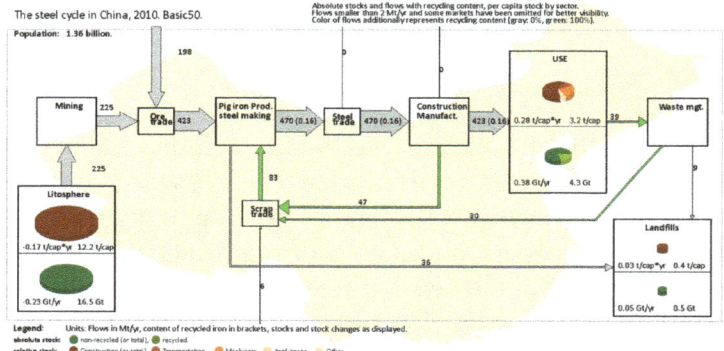

A typical MFA system with quantification.

Process Balance

One of the main purposes of MFA is to quantify the metabolism of the elements of the system. Unlike economic accounting, MFA also covers non-economic waste flows, emissions to the environment, and non-market natural resources.

The process balance is a first order physical principle that turns MFA into a powerful accounting and analysis tool. The nature of the processes in the system determine which balances apply. For a process 'oil refinery', for example, one can establish a mass balance for each chemical element, while this is not possible for a nuclear power station. A car manufacturing plant respects the balance for steel, but a steel mill does not.

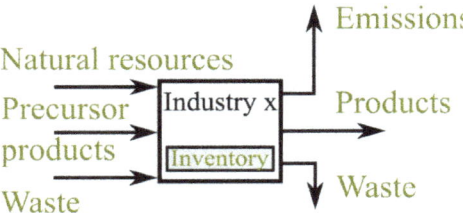

Model of an industrial process in economic accounting (top) and in physical accounting (bottom).

When quantifying MFA systems either by measurements or from statistical data, mass and other process balances have to be checked to ensure the correctness of the quantification and to reveal possible data inconsistencies or even misconceptions in the system such as the omission of a flow or a process. Conflicting information can be reconciled using data validation and reconciliation, and the STAN-software offers basic reconciliation functionality that is suitable for many MFA application.

Examples of MFA Applications on Different Spatial and Temporal Scales

MFA studies are conducted on various spatial and temporal scales and for a variety of elements, substances, and goods. They cover a wide range of process chains and material cycles. Several examples:

- *MFA on a national or regional scale* (also referred to as material flow accounting): In this type of study, the material exchanges between an economy and the natural environment are analyzed. Several indicators are calculated in order to assess the level of resource intensity of the system.
- *Corporate material flow analysis*, or MFA along an *industrial supply chain* involves a number of companies: The goal of material flow analysis within a company is to quantify and then optimize the production processes so that materials and energy are used more efficiently manner, e.g., by recycling and waste reduction. Companies can use the results obtained by Material Flow Cost Accounting to reduce their operational costs and improve environmental performance.
- In the *life cycle of a product*: The life cycle inventory, whose compilation is at the core of life cycle assessment, follows the MFA methodology as it is based on an explicit system definition and process balances.

Historical Development

- Mass balance or the conservation of matter has been postulated already in ancient Greece, and it was introduced into modern chemistry by Antoine Lavoisier, from where it found its way to chemical engineering and finally to environmental science.

- Other seminal contributions were made by Sanctorius and Theodor Weyl.

- Dennis Meadows made a wide audience aware of the physical foundation of the economy when he co-authored the bestseller *Limits to Growth* in 1971. Meadows et al. based their predictions on an analysis of resource stocks;

- The methodology of MFA was developed during the 80s and 90s. Development happened simultaneously in different research groups. Central publications on the MFA methodology include Baccini and Bader (1996), Brunner and Rechberger (2004), Baccini and Brunner (2012), and van der Voet et al. (2002).

- Friedrich Schmidt-Bleek, who worked at the Wuppertal Institute, developed the MFA-related concept of Material Input Per Service unit (MIPS).

- The UNEP Resource Panel was set up in 2007 by the United Nations Environment Program. In analogy to the Intergovernmental Panel on Climate Change (IPCC) it brings together experts from many disciplines and institutions to review the current state of research on societal metabolism and to communicate the latest findings to policymakers and stakeholders.

Recent Development

- MFA concepts have been or are being incorporated in national accounts in several countries and regions such as the EU and Japan. MFA is also used in the System of Integrated Environmental and Economic Accounting.

- Several international conferences or other meetings provide a platform for researchers and policymakers to meet and exchange results and ideas, including the World Resources Forum, a bi-annual international conference on material flow analysis and sustainable development.

- MFA-IO is an approach to establish an MFA system of the whole economy using monetary and physical Input-Output tables.

- The Sustainable Europe Research Institute (SERI) in Vienna, Austria, has developed a database called materialflows.net.

- Dynamic MFA aims for long-term quantification of MFA systems and uses historic development patterns of physical stocks and flows to create robust scenarios for the years and decades to come.

- Japan has developed into a hotspot for MFA research. The country has scarce mineral resources and therefore depends on imports of energy carriers, ores, and other raw materials. The Japanese government fosters research on material cycles and also inaugurated the 3-R concept.

Conducting a State-of-the-art MFA

A state-of-the-art MFA consists of the following steps:

- Establish an explicit system definition: Specify the system boundary with geographical and temporal scope, processes (can contain stocks), and flows. Specify the material for which the system is to be quantified (product, substance, or indicator element). Make sure that each stock is associated with a process and that each flow connects one process to another. Flows can also begin or end outside the system boundary.
- Define and name the system variables. The system variables include: All stocks within the processes, all flows between processes, and all flows coming from outside or going to outside the system boundaries. Sometimes, stocks are not considered and only the net stock changes are of interest. For each variable it must be clear whether it is a stock or a flow, and this distinction needs to be reflected in the names and in the mathematical symbols chosen.
- Quantify the system variables by linking them to literature, measurement, or modelled data.
- Performa a mass balance check for all processes and the system as a whole.
- Optional: Visualise your system by using the box-and-arrow scheme shown above or by using Sankey diagrams.
- Document the MFA by reporting the explicit system definition along with the list of quantified system variables and the mass balance checks.

The Difference Between Material and Substance Flow Analysis

While the term 'substance' in 'substance flow analysis (SFA) always refers to chemical substances, the term 'material' in 'material flow analysis (MFA)' has a much wider scope. According to Brunner and Rechberger the term 'material' comprises substances AND goods, and the reason for this wide scope is the wish to apply MFA not only to chemical elements or substances but also to materials like steel, timber, or products like cars or buildings. It is thus possible to conduct an MFA for the passenger vehicle fleet by recording the vehicles entering and leaving the use phase.

Relation to Other Methods

MFA is complementary to the other core industrial ecology methods life cycle assessment (LCA) and input-output (I/O) models. Some overlaps between the different methods exist as they all share the system approach and to some extent the mass balance principle. The methods mainly differ in purpose, scope, and data requirements.

MFA studies often cover the entire cycle (mining, production, manufacturing, use, waste handling) of a certain substance within a given geographical boundary and time frame. Material stocks are explicit in MFA, which makes this method suitable for studies involving resource scarcity and recycling from old scrap. The common use of time series (dynamic modelling) and lifetime models makes MFA a suitable tool for assessing long-term trends in material use.

- Compared to I/O analyses, the number of processes considered in MFA systems is usually much lower. On the other hand, mass balance ensures that flows of by-products or waste are not overlooked in MFA studies, whereas in I/O tables these flows are often not included due to their lack of economic value. Physical I/O models are much less common than economic tables. Material stocks are not covered by IO analysis, only the addition to stock can be included in form of capital accumulation.
- Life cycle inventories record the demand for many different materials associated with individual products, whereas MFA studies typically focus on a single material used in many different products.

Cleaner Production

Cleaner production is a preventive, company-specific environmental protection initiative. It is intended to minimize waste and emissions and maximize product output. By analysing the flow of materials and energy in a company, one tries to identify options to minimize waste and emissions out of industrial processes through source reduction strategies. Improvements of organisation and technology help to reduce or suggest better choices in use of materials and energy, and to avoid waste, waste water generation, and gaseous emissions, and also waste heat and noise.

The concept was developed during the preparation of the Rio Summit as a programme of UNEP (United Nations Environmental Programme) and UNIDO (United Nations Industrial Development Organization) under the leadership of Jacqueline Aloisi de Larderel, the former Assistant Executive Director of UNEP. The programme was meant to reduce the environmental impact of industry. It built on ideas used by 3M in its 3P programme (pollution prevention pays). It has found more international support than all other comparable programmes. The programme idea was described "to assist developing nations in leapfrogging from pollution to less pollution, using available technologies". Starting from the simple idea to produce with less waste Cleaner Production was developed into a concept to increase the resource efficiency of production in general. UNIDO has been operating National Cleaner Production Centers and Programmes (NCPCs/NCPPs) with centres in Latin America, Africa, Asia and Europe.

In the US, the term pollution prevention is more commonly used for cleaner production.

Examples for cleaner production options are:

- Documentation of consumption (as a basic analysis of material and energy flows, e. g. with a Sankey diagram)
- Use of indicators and controlling (to identify losses from poor planning, poor education and training, mistakes)
- Substitution of raw materials and auxiliary materials (especially renewable materials and energy)
- Increase of useful life of auxiliary materials and process liquids (by avoiding drag in, drag out, contamination)
- Improved control and automatisation
- Reuse of waste (internal or external)
- New, low waste processes and technologies

One of the first European initiatives in cleaner production was started in Austria in 1992 by the BMVIT (Bundesministerium für Verkehr, Innovation und Technologie). This resulted in two initiatives: "Prepare" and EcoProfit.

The "PIUS" initiative was founded in Germany in 1999. Since 1994, the United Nations Industrial Development Organization operates the National Cleaner Production Centre Programme with centres in Central America, South America, Africa, Asia, and Europe.

Plant Simulation

Plant Simulation is a computer application developed by Siemens PLM Software for modeling, simulating, analyzing, visualizing and optimizing production systems and processes, the flow of materials and logistic operations. Using Tecnomatix Plant Simulation, users can optimize material flow, resource utilization and logistics for all levels of plant planning from global production facilities, through local plants, to specific lines. Within the *Plant Design und Optimization Solution* the software portfolio, to which Plant Simulation belongs, is — together with the products of the Digital Factory and of Digital Manufacturing — part of the Product Lifecycle Management Software (PLM). The application allows comparing complex production alternatives, including the immanent process logic, by means of computer simulations. Plant Simulation is used by individual production planners as well as by multi-national enterprises, primarily to strategically plan layout, control logic and dimensions of large, complex production investments. It is one of the major products that dominate that market space.

Product Description

Plant Simulation is a Material flow simulation Software (Discrete Event Simulation; DES Software). Using simulation, complex and dynamic enterprise workflows are evaluated to arrive at mathematically safeguarded entrepreneurial decisions. The Computer model allows the user to execute experiments and to run through 'what if scenarios' without either having to experiment with the real production environment or, when applied within the planning phase, long before the real system exists. In general the Material flow analysis is used when discrete production processes are running. These processes are characterized by non-steady material flows, which means that the part is either there or not there, the shift takes place or does not take place, the machine works without errors or reports a failure. These processes resist simple mathematical descriptions and derivations due to numerous dependencies. Before powerful computers were available, most problems of material flow simulation have been solved by means of queuing theory and operations research methods. In most cases the solutions resulting from these calculations were hard to understand and were marked by a large number of boundary conditions and restrictions which were hard to abide by in reality.

Languages

Plant Simulation is available in English, German, Japanese, Hungarian, Russian and Chinese. The user can create individual Dialog boxes using double-byte characters and offering individual parameterizations. The user can switch between the available languages.

Special Features

- Object-oriented programming with

    o Inheritance: Users create libraries with their own objects, which can be reused. As opposed to a copy, any change to an object class within the library is propagated to any of the derived objects (children).

    o Polymorphism: Classes can be derived and derived methods can be redefined. This enables users to build complex models faster, easier and with a clearer structure.

    o Hierarchy: Complex structures can be created very clearly on several (logic) layers. This facilitates a Top-down and bottom-up design approach.

- Openness for importing data from other systems, such as Access or Oracle data bases, Excel worksheets or from SAP.

- Integration: Plant Simulation is part of the Digital factory and supports
  - importing data from PLM systems or be used during
  - Virtual Commissioning
  - taking over layout data from Autocad, Microstation, FactoryCAD, etc. directly into the simulation.

- Provides comprehensible analysis tools for detecting bottlenecks (Bottleneck Analyzer), for tracking the flow of materials (Sankey diagrams) or for detecting over-dimensioned resources (Chart Wizard).

- Provides integrated optimization tools:
  - The Experiment Manager automatically creates scenarios or evaluates dependencies between two input parameters.
  - Genetic algorithms search large solution spaces.
  - Neural networks show the connection between input and output parameters and can be used for forecasting.

- Data analysis: Detection of dependencies, Regression analysis, best fitting function etc.

Scope of Application

Calculation of Enterprise Characteristics

Goal:

- Detect and show problems which might otherwise cause costs and time-intensive correction measures during the ramp-up phase.

- Offer mathematically calculated key performance indicators (KPI) instead of expert's "gut feelings."

- Reduce investment costs for production lines without endangering the required output quantities.

- Optimize the performance of existing production lines.

- Incorporate machine failures, availabilities (MTTR, MTBF) when calculating throughput numbers and utilization.

Visualization

Plant Simulation can display production sequences in 2D and in 3D. The 3D display is especially helpful as a sales tool or for in-house communication of planned measures. In addition it allows to present the entire system concept within a virtual, interactive, immersive environment to non-simulation experts. The 3D engine is based on the industry standard JT format. CAD applications such as NX, Solid Edge can export models in this format. The 3D data files can be imported in the JT format '.jt' by using Drag-and-drop.

Used in

Plant Simulation is used in most industries. Especially in the

- Automotive industry Automotive Industry Workgroup Material Flow Simulation
- Automotive suppliers
- Aerospace
- Plant manufacturing
- Mechanical engineering
- Process industry
- Electronics industry
- Consumer packaged goods industry
- Airports
- Logistics companies (transport logistics, storage logistics and production logistics)
- High bay warehouse suppliers, suppliers of automated guided vehicle systems and electric overhead monorail systems
- Consulting houses and service providers
- Shipyards Simulation Cooperation in the Maritime Industries; SimCoMar is an interest group of shipyards and suppliers, universities and institutions engaged in the simulation of shipbuilding
- Harbors, especially container terminals

Lately material flow simulation gains growing importance through the increasing use for considering the sustainability of industrial production processes. Here the characteristics of sustainable manufacturing are simulated and analyzed beforehand and then integrated into the investment decision process. Plant Simulation is also used for research and development purposes at a great number of universities and universities of applied science.

Application History

Year	Company	Product name
1986	The Fraunhofer Society for Factory Operation and Automation develops an object-oriented, hierarchical simulation program for the Apple Macintosh	SIMPLE Mac for Apple Macintosh
1990	AIS (Angewande Informations Systeme) founded	SIMPLE++ (Simulation in Produktion Logistik and Engineering)
1991	AIS renamed to AESOP (Angewande EDV-Systeme zur optimierten Planung)	SIMPLE++ (Simulation in Produktion Logistik und Engineering)
1997	AESOP acquired by Tecnomatix Ltd.	2000 SIMPLE++ renamed to eM-Plant
2004	Tecnomatix Ltd. acquired by UGS Corporation	2005 eM-Plant renamed to Tecnomatix Plant Simulation
2007	UGS Corporation acquired by Siemens AG	Tecnomatix Plant Simulation

AnyLogic

AnyLogic is a multimethod simulation modeling tool developed by The AnyLogic Company (former XJ Technologies). It supports agent-based, discrete event, and system dynamics simulation methodologies. AnyLogic PLE edition is available for free for self-educational and educational purposes.

History of AnyLogic

In the beginning of the 1990s there was a big interest in the mathematical approach to modeling and simulation of parallel processes. This approach may be applied to the analysis of correctness of parallel and distributed programs. The Distributed Computer Network (DCN) research group at Saint Petersburg Technical University developed such a software system for the analysis of program correctness; the new tool was named COVERS (Concurrent Verification and Simulation). This system allowed graphical modeling notation for system structure and behavior. The tool was applied for the research granted by Hewlett Packard.

In 1998 the success of this research inspired the DCN laboratory to organize a company with a mission to develop a new age simulation software. The emphasis in the development was placed on applied methods: simulation, performance analysis, behavior of stochastic systems, optimization and visualization. New software released in 2000 was based on the latest advantages of information technologies: an object-oriented approach, elements of the UML standard, the use of Java, a modern GUI, etc.

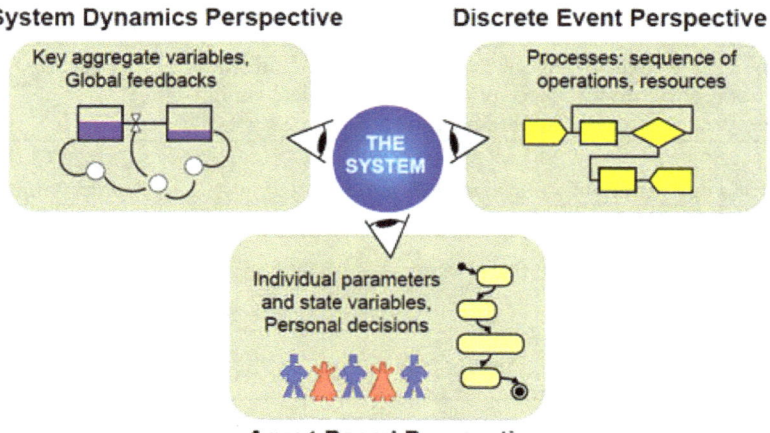

Three business simulation approaches

The tool was named AnyLogic, because it supported all three well-known modeling approaches:

- System dynamics,
- Discrete event simulation,
- Agent-based modeling.

+ Any combination of these approaches within a single model. The first version of AnyLogic was AnyLogic 4, because the numbering continues the numbering of COVERS 3.0.

A big step was taken in 2003, when AnyLogic 5 was released. It was focused on business simulation in the following domains:

- Market and Competition,
- Healthcare,
- Manufacturing,
- Supply Chain,
- Logistics,
- Retail,

- Business Processes,
- Social and Ecosystem Dynamics,
- Defense,
- Project and Asset Management,
- IT Infrastructure,
- Pedestrian Dynamics and Traffic simulation,
- Aerospace.
- Photovoltaics

AnyLogic 7, was released in 2014. Being the biggest release for 7 years, it featured many significant updates aimed at simplifying model building, including enhanced support for multimethod modeling, decreased need for coding, renewed libraries, and other usability improvements. AnyLogic 7.1, also released in 2014, included the new GIS implementation in the software: in addition to shapefile-based maps, AnyLogic started to support tile maps from free online providers, including OpenStreetMap.

2015 marked the release of AnyLogic 7.2 with the built-in database and the Fluid Library. The free Personal Learining Edition (PLE) was also introduced in 2015.

The new Road Traffic Library was introduced in 2016 with AnyLogic 7.3.

AnyLogic 8 was released in 2017. Beginning with Version 8.0, the AnyLogic model development environment was integrated with AnyLogic Cloud, a web service for simulation analytics.

The platform for AnyLogic 8 model development environment is Eclipse. AnyLogic is a cross-platform simulation software as far as it works on Windows, Mac OS and Linux.

AnyLogic and Java

AnyLogic includes a graphical modeling language and also allows the user to extend simulation models with Java code. The Java nature of AnyLogic lends itself to custom model extensions via Java coding as well as the creation of Java applets which can be opened with any standard browser. These applets make AnyLogic models very easy to share or place on websites. In addition to Java applets the Professional version allows for the creation of Java runtime applications which can be distributed to users. These pure Java applications can be a base for decision support tools. It should be noted, however, that most internet browsers have disabled Java applet support due to security concerns.

Multimethod Simulation Modeling

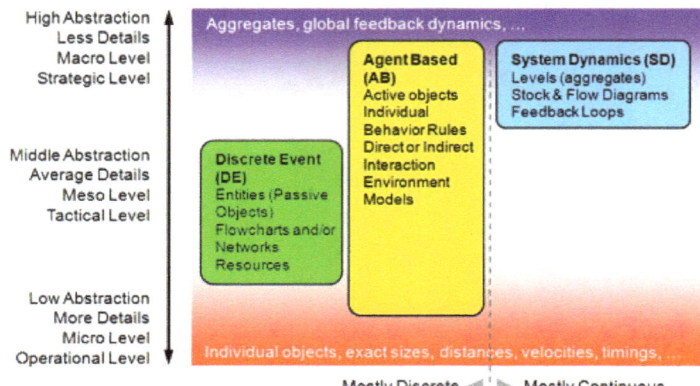

How simulation approaches correspond to the level of abstraction

AnyLogic models can be based on any of the main simulation modeling paradigms: discrete event or process-centric (DE), systems dynamics (SD), and agent-based (AB).

System dynamics and discrete event are traditional simulation approaches, agent based is a newer one. Technically, system dynamics approach deals mostly with continuous processes whereas "discrete event" (by which we mean all descendants of GPSS also known as process-centric simulation approach) and agent based models work mostly in discrete time, i.e. jump from one event to another.

System dynamics and discrete event simulation historically have been taught at universities to very different groups of students, namely management and economy, industrial and operation research engineers. As a result, there are two distinct practitioners' communities that never talk to each other.

Agent based modeling until recently has been mostly a purely academic topic. However, the increasing demand for global business optimization caused leading modelers looking at combined approaches to gain a deeper insight into complex interdependent processes having very different natures.

How modeling approaches correspond to the abstraction levels? System dynamics dealing with aggregates is obviously used at the highest abstraction level. Discrete event modeling is used at low to middle abstraction. As for agent based modeling, this technology is used across all abstraction levels, and agent may model objects of very diverse nature and scale: at the "physical" level agents may be e.g. pedestrians or cars or robots, at the middle level – customers, at the highest level – competing companies.

AnyLogic allows the modeler to combine these simulation approaches within the same model. There is no fixed hierarchy. So, as an example, one could create a model of the package shipping industry where carriers are modeled as agents acting/reacting independently whereas the inner workings of their transport and infrastructure networks could be modeled with discrete event simulation. Similarly, one can model consumers as agents whose aggregate behavior feed a systems dynamics model capturing flows such as revenues or costs which do not need to be tied to individual agents. This mixed language approach is directly applicable to a wide variety of complex modeling problems that may be modeled via any one approach albeit with compromises.

Features

Simulation Language

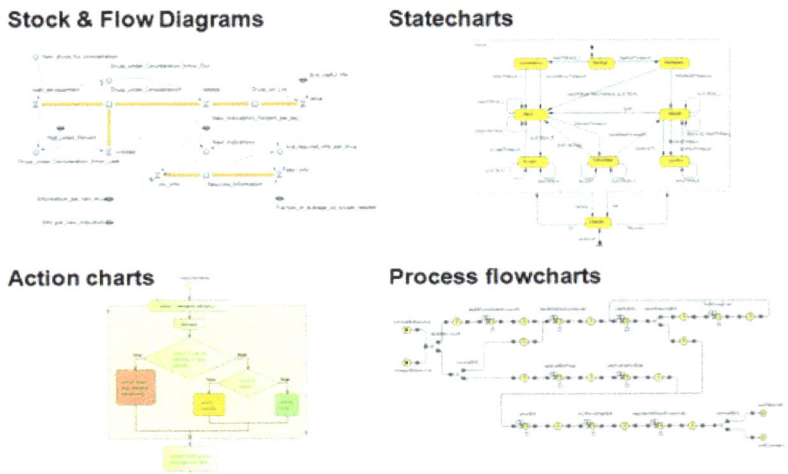

Simulation language constructions provided by AnyLogic

The AnyLogic simulation language consists of following items:

- Stock & Flow Diagrams are used for System Dynamics modeling.

- Statecharts are used mostly in Agent Based modeling to define agent behavior. They are also often used in Discrete Event modeling, e.g. to simulate machine failure.

- Action charts are used to define algorithms. They may be used in Discrete Event modeling, e.g. for call routing, or in Agent Based modeling, e.g. for agent decision logic.

- Process flowcharts are the basic construction used to define process in Discrete Event modeling. Looking at this flowchart you may see why Discrete Event style is often called Process Centric.

The language also includes: low level modeling constructions (variables, equations, parameters, events etc.), presentation shapes (lines, polylines, ovals etc.), analysis facilities (datasets, histograms, plots), connectivity tools, standard images, and experiments frameworks.

AnyLogic Libraries

AnyLogic includes the following standard libraries:

- The Process Modeling Library is designed to support DE simulation in Manufacturing, Supply Chain, Logistics and Healthcare areas. Using the Process Modeling Library objects you can model real-world systems in terms of entities (transactions, customers, products, parts, vehicles, etc.), processes (sequences of operations typically involving queues, delays, resource utilization), and resources. The processes are specified in the form of flowcharts. The Process Modeling Library is a successor of the Enterprise Library from AnyLogic 6, which is also available in AnyLogic 7.

- The Pedestrian Library is dedicated to simulating pedestrian flows in a physical environment. It allows you to create models of pedestrian-intensive buildings (like subway stations, security checks etc.) or streets (large numbers of pedestrians). Models support statistics collection on pedestrian density in different areas. This ensures acceptable performance of service points with a hypothetical load, estimates lengths of stay in specific areas, and detects potential problems with interior geometry – such as the effect of adding too many obstacles - and other applications. In models created with the Pedestrian Library, pedestrians move in continuous space, reacting to different kinds of obstacles (walls, different kinds of areas), as well as other pedestrians. Pedestrians are simulated as interacting agents with complex behavior, but the AnyLogic Pedestrian Library provides a higher level interface for faster creation of pedestrian models in the style of flowcharts.

- The Rail Library supports modeling, simulating, and visualizing operations of a rail yard of any complexity and scale. The rail yard models can be combined with discrete event or agent based models related to: loading and unloading, resource allocation, maintenance, business processes, and other transportation activities.

- The Fluid Library allows the user to model storage and transfer of fluids, bulk goods, or large amounts of discrete items, which are not desirable to model as separate objects. The library includes blocks such as tank, pipeline, valve, and objects for routing, merging, and diverging the flow. To improve model execution speed, the Fluid Library uses a linear programming solver. The library is designed to improve AnyLogic use in manufacturing, oil, gas, and mining industries. The user can simulate oil pipes and tanks, ore, coal conveyors, and production processes where liquids or bulk materials are involved, for example, in concrete manufacturing.

- The Road Traffic Library allows users to simulate vehicle traffic on roads. The library supports detailed, physical level modeling of vehicle movement. Each vehicle represents an agent that can have its own behavioral patterns inside. The library allows users to simulate vehicle movement on roads, taking into account driving regulations, traffic lights, pedestrian crossings, priorities at junctions, parking lots, and public transport movements. The library is suitable for modeling highway traffic, street traffic, on-site transportation at manufacturing sites, or any other systems with vehicles, roads, and lanes. A special traffic density tool is included to help analyze road network loads.

Besides these standard libraries users can create their own ones and distribute them.

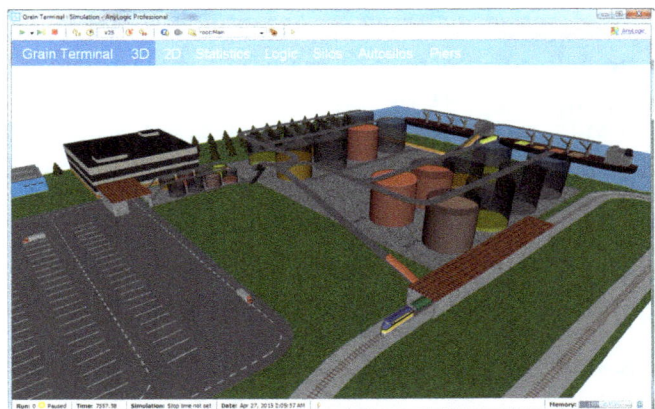

Grain Terminal Model

Model Animation

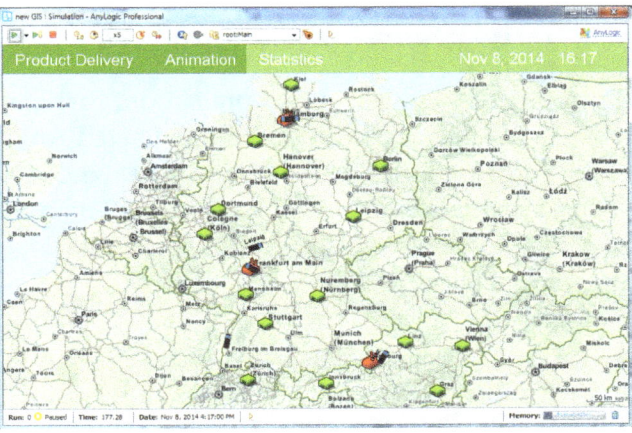

GIS-Based Supply Chain Simulation Model

AnyLogic supports interactive 2D and 3D animation.

AnyLogic allows users to import CAD drawings as DXF files, and then visualize models on top of them. This feature can be used for animating processes inside objects like factories, warehouses, hospitals, etc. This functionality is mostly used in Discrete Event (process-based) models in manufacturing, healthcare, civil engineering, and construction. AnyLogic software also supports 3D animation and includes a collection of ready-to-use 3D objects for animation related to different industries, including buildings, road, rail, maritime, transport, energy, warehouse, hospital, equipment, airport-related items, supermarket-related items, cranes, and other objects.

Models can include custom UI for users to configure experiments and change input data.

Geospatial Models, GIS Integration

AnyLogic models can use maps as a layout, which is often required by supply chains, logistics, and transportation industries. AnyLogic software supports the traditional shapefile-based map standard, SHP by Esri. In addition, AnyLogic supports tile maps from free online providers, including OpenStreetMap. Tile maps allow the modeler to use map data in models and to automatically create geospatial routes for agents. The main tile map features in AnyLogic include:

- The model can access all of the data stored along with online-based maps: cities, regions, road networks, and objects (hospitals, schools, bus stops, etc.).

- Agents can be placed in specified points on the map, and moved along existing roads or routes.

- Users can create the required elements inside the model using the built-in search.

Model Integration with Other IT-infrastructure

An AnyLogic model can be exported as a Java application, that can be run separately, or integrated with other software. As an option, an exported AnyLogic model can be built into other pieces of software and work as an additional module to ERP, MRP, and TMS systems. Another typical use is integration of an AnyLogic model with TXT, MS Excel, or MS Access files and databases (MS SQL, My SQL, Oracle, etc.). Also, Anylogic models include their own databases based on HSQLDB.

AnyLogic Cloud

AnyLogic Cloud is a web service for simulation analytics. It allows users to store, access, run, and share simulation models online, as well as analyze experiment results.

Using AnyLogic model development environment, developers can upload their models to AnyLogic Cloud and set up sharable web dashboards to work with models online. These dashboards can contain configurable input parameters and output data in the form of charts and graphs. Model users can set input data on the dashboard screen, run the model, and analyze the output.

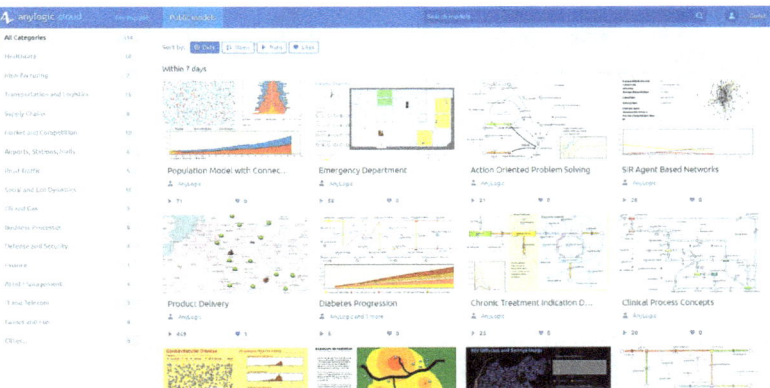

View of AnyLogic Cloud public model library

AnyLogic Cloud allows users to run models using web browsers, on desktop computers and mobile devices, with the model being executed on the server side. Multiple run experiments are performed using several nodes. The results of all executed experiments are stored in the database and can be immediately accessed. Models can be run both with and without HTML5-based interactive animation.

Developers can choose whether they want their models to be private or publicly available in the model library, which includes models from other AnyLogic users.

Free Educational Version

Since 2015, AnyLogic Personal Learning Edition (PLE) is available for free for the purposes of education and self-education. The PLE license is perpetual, but created models are limited in size.

For public research in educational institutions, users can obtain a discounted University Researcher license, which does not limit model size and has a lot of the functionality of a Professional license.

AnyLogistix Supply Chain Optimization Software

AnyLogic does not include a specific library for supply chain simulation, as The AnyLogic Company converted its development efforts for this domain in a separate software tool – anyLogistix. This spin-off product was introduced in 2014 as AnyLogic Logistics Network Manager and was renamed anyLogistix in 2015.

anyLogistix is based on the AnyLogic engine, GIS, and the new industry-oriented GUI. It also includes algorithms and techniques specific for supply chain design and optimization. anyLogistix is fully integrated with AnyLogic, for instance, AnyLogic can be used for customization of objects inside anyLogistix, including warehouses, production sites, suppliers, inventory, sourcing, and transportation policies. In addition, anyLogistix uses IBM ILOG CPLEX Optimization Studio for optimization problems.

Supply Chain Management

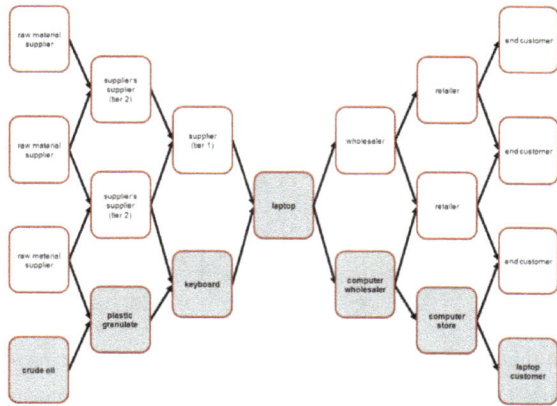

Supply chain management field of operations: complex and dynamic supply- and demand-networks. (cf. Wieland/Wallenburg, 2011)

In commerce, supply chain management (SCM), the management of the flow of goods and services, involves the movement and storage of raw materials, of work-in-process inventory, and of finished goods from point of origin to point of consumption. Interconnected or interlinked networks, channels and node businesses combine in the provision of products and services required by end customers in a supply chain. Supply-chain management has been defined as the "design, planning, execution, control, and monitoring of supply chain activities with the objective of creating net value, building a competitive infrastructure, leveraging worldwide logistics, synchronizing supply with demand and measuring performance globally."

SCM practice draws heavily from the areas of industrial engineering, systems engineering, operations management, logistics, procurement, information technology, and marketing and strives for an integrated approach. Marketing channels play an important role in supply chain management Current research in supply chain management is concerned with topics related to sustainability and risk management, among others, whereas the "people dimension" of SCM, ethical issues, internal integration, transparency/visibility, and human capital/talent management are topics that have, so far, been underrepresented on the research agenda.

Origin of the Term and Definitions

In 1982, Keith Oliver, a consultant at Booz Allen Hamilton (now Strategy&), introduced the term "supply chain management" to the public domain in an interview for the Financial Times.

In the mid-1990s, more than a decade later, the term "supply chain management" gained currency when a flurry of articles and books came out on the subject. Supply chains were originally defined as encompassing all activities associated with the flow and transformation of goods from raw materials through to the end user, as well as the associated information flows. Supply chain management was then further defined as the integration of supply chain activities through improved supply chain relationships to achieve a competitive advantage.

In the late 1990s, "supply chain management" (SCM) rose to prominence as a management buzzword, and operations managers began to use it in their titles with increasing regularity.

Other commonly accepted definitions of supply chain management include:

- The management of upstream and downstream value-added flows of materials, final goods, and related information among suppliers, company, resellers, and final consumers.

- The systematic, strategic coordination of traditional business functions and tactics across all business functions within a particular company and across businesses within the supply chain, for the purposes of improving the long-term performance of the individual companies and the supply chain as a whole

- A customer-focused definition is given by Hines (2004:p76): "Supply chain strategies require a total systems view of the links in the chain that work together efficiently to create customer satisfaction at the end point of delivery to the consumer. As a consequence, costs must be lowered throughout the chain by driving out unnecessary expenses, movements, and handling. The main focus is turned to efficiency and added value, or the end-user's perception of value. Efficiency must be increased, and bottlenecks removed. The measurement of performance focuses on total system efficiency and the equitable monetary reward distribution to those within the supply chain. The supply chain system must be responsive to customer requirements."

- The integration of key business processes across the supply chain for the purpose of creating value for customers and stakeholders (Lambert, 2008)

- According to the Council of Supply Chain Management Professionals (CSCMP), supply chain management encompasses the planning and management of all activities involved in sourcing, procurement, conversion, and logistics management. It also includes coordination and collaboration with channel partners, which may be suppliers, intermediaries, third-party service providers, or customers. Supply chain management integrates supply and demand management within and across companies. More recently, the loosely coupled, self-organizing network of businesses that cooperate to provide product and service offerings has been called the *Extended Enterprise*.

A supply chain, as opposed to supply chain management, is a set of organizations directly linked by one or more upstream and downstream flows of products, services, finances, or information from a source to a customer. Supply chain management is the management of such a chain.

Supply chain management software includes tools or modules used to execute supply chain transactions, manage supplier relationships, and control associated business processes.

Supply chain event management (SCEM) considers all possible events and factors that can disrupt a supply chain. With SCEM, possible scenarios can be created and solutions devised.

In many cases the supply chain includes the collection of goods after consumer use for recycling. Including third-party logistics or other gathering agencies as part of the RM re-patriation process is a way of illustrating the new endgame strategy.

Functions

Supply chain management is a cross-functional approach that includes managing the movement of raw materials into an organization, certain aspects of the internal processing of materials into finished goods, and the movement of finished goods out of the organization and toward the end consumer. As organizations strive to focus on core competencies and become more flexible, they reduce their ownership of raw materials sources and distribution channels. These functions are increasingly being outsourced to other firms that can perform the activities better or more cost effectively. The effect is to increase the number of organizations involved in satisfying customer demand, while reducing managerial control of daily logistics operations. Less control and more supply chain partners lead to the creation of the concept of supply chain management. The purpose of supply chain management is to improve trust and collaboration among supply chain partners, thus improving inventory visibility and the velocity of inventory movement.

Importance

Organizations increasingly find that they must rely on effective supply chains, or networks, to compete in the global market and networked economy. In Peter Drucker's (1998) new management paradigms, this concept of business relationships extends beyond traditional enterprise boundaries and seeks to organize entire business processes throughout a value chain of multiple companies.

In recent decades, globalization, outsourcing, and information technology have enabled many organizations, such as Dell and Hewlett Packard, to successfully operate collaborative supply networks in which each specialized business partner focuses on only a few key strategic activities (Scott, 1993). This inter-organisational supply network can be acknowledged as a new form of organisation. However, with the complicated interactions among the players, the network structure fits neither "market" nor "hierarchy" categories

(Powell, 1990). It is not clear what kind of performance impacts different supply network structures could have on firms, and little is known about the coordination conditions and trade-offs that may exist among the players. From a systems perspective, a complex network structure can be decomposed into individual component firms (Zhang and Dilts, 2004). Traditionally, companies in a supply network concentrate on the inputs and outputs of the processes, with little concern for the internal management working of other individual players. Therefore, the choice of an internal management control structure is known to impact local firm performance (Mintzberg, 1979).

In the 21st century, changes in the business environment have contributed to the development of supply chain networks. First, as an outcome of globalization and the proliferation of multinational companies, joint ventures, strategic alliances, and business partnerships, significant success factors were identified, complementing the earlier "just-in-time", lean manufacturing, and agile manufacturing practices. Second, technological changes, particularly the dramatic fall in communication costs (a significant component of transaction costs), have led to changes in coordination among the members of the supply chain network (Coase, 1998).

Many researchers have recognized supply network structures as a new organisational form, using terms such as "Keiretsu", "Extended Enterprise", "Virtual Corporation", "Global Production Network", and "Next Generation Manufacturing System". In general, such a structure can be defined as "a group of semi-independent organisations, each with their capabilities, which collaborate in ever-changing constellations to serve one or more markets in order to achieve some business goal specific to that collaboration" (Akkermans, 2001).

Supply chain management is also important for organizational learning. Firms with geographically more extensive supply chains connecting diverse trading cliques tend to become more innovative and productive.

The security management system for supply chains is described in ISO/IEC 28000 and ISO/IEC 28001 and related standards published jointly by the ISO and the IEC. Supply Chain Management draws heavily from the areas of operations management, logistics, procurement, and information technology, and strives for an integrated approach.

Historical Developments

Six major movements can be observed in the evolution of supply chain management studies: creation, integration, and globalization (Movahedi et al., 2009), specialization phases one and two, and SCM 2.0.

Creation Era

The term "supply chain management" was first coined by Keith Oliver in 1982. However, the concept of a supply chain in management was of great importance long before, in the early 20th century, especially with the creation of the assembly line. The characteristics of this era of supply chain management include the need for large-scale changes, re-engineering, downsizing driven by cost reduction programs, and widespread attention to Japanese management practices. However, the term became widely adopted after the publication of the seminal book *Introduction to Supply Chain Management* in 1999 by Robert B. Handfield and Ernest L. Nichols, Jr., which published over 25,000 copies and was translated into Japanese, Korean, Chinese, and Russian.

Integration Era

This era of supply chain management studies was highlighted with the development of electronic data interchange (EDI) systems in the 1960s, and developed through the 1990s by the introduction of enterprise resource planning (ERP) systems. This era has continued to develop into the 21st century with the expansion of Internet-based collaborative systems. This era of supply chain evolution is characterized by both increasing value added and cost reductions through integration.

A supply chain can be classified as a stage 1, 2 or 3 network. In a stage 1–type supply chain, systems such as production, storage, distribution, and material control are not linked and are independent of each other. In a stage 2 supply chain, these are integrated under one plan and is ERP enabled. A stage 3 supply chain is one that achieves vertical integration with upstream suppliers and downstream customers. An example of this kind of supply chain is Tesco.

Globalization Era

The third movement of supply chain management development, the globalization era, can be characterized by the attention given to global systems of supplier relationships and the expansion of supply chains beyond national boundaries and into other continents. Although the use of global sources in organisations' supply chains can be traced back several decades (e.g., in the oil industry), it was not until the late 1980s that a considerable number of organizations started to integrate global sources into their core business. This era is characterized by the globalization of supply chain management in organizations with the goal of increasing their competitive advantage, adding value, and reducing costs through global sourcing.

Specialization Era (phase I): Outsourced Manufacturing and Distribution

In the 1990s, companies began to focus on "core competencies" and specialization. They abandoned vertical integration, sold off non-core operations, and outsourced those functions to other companies. This changed management requirements, by extending the supply chain beyond the company walls and distributing management across specialized supply chain partnerships.

This transition also refocused the fundamental perspectives of each organization. Original equipment manufacturers (OEMs) became brand owners that required visibility deep into their supply base. They had to control the entire supply chain from above, instead of from within. Contract manufacturers had to manage bills of material with different part-numbering schemes from multiple OEMs and support customer requests for work-in-process visibility and vendor-managed inventory (VMI).

The specialization model creates manufacturing and distribution networks composed of several individual supply chains specific to producers, suppliers, and customers that work together to design, manufacture, distribute, market, sell, and service a product. This set of partners may change according to a given market, region, or channel, resulting in a proliferation of trading partner environments, each with its own unique characteristics and demands.

Specialization Era (phase II): Supply Chain Management as a Service

Specialization within the supply chain began in the 1980s with the inception of transportation brokerages, warehouse management (storage and inventory), and non-asset-based carriers, and has matured beyond transportation and logistics into aspects of supply planning, collaboration, execution, and performance management.

Market forces sometimes demand rapid changes from suppliers, logistics providers, locations, or customers in their role as components of supply chain networks. This variability has significant effects on supply chain infrastructure, from the foundation layers of establishing and managing electronic communication between trading partners, to more complex requirements such as the configuration of processes and work flows that are essential to the management of the network itself.

Supply chain specialization enables companies to improve their overall competencies in the same way that outsourced manufacturing and distribution has done; it allows them to focus on their core competencies and assemble networks of specific, best-in-class partners to contribute to the overall value chain itself, thereby increasing overall performance and efficiency. The ability to quickly obtain and deploy this domain-specific supply chain expertise without developing and maintaining an entirely unique and complex competency in house is a leading reason why supply chain specialization is gaining popularity.

Outsourced technology hosting for supply chain solutions debuted in the late 1990s and has taken root primarily in transportation and collaboration categories. This has progressed from the application service provider (ASP) model from roughly 1998 through 2003, to the on-demand model from approximately 2003 through 2006, to the software as a service (SaaS) model currently in focus today.

Supply Chain Management 2.0 (SCM 2.0)

Building on globalization and specialization, the term "SCM 2.0" has been coined to describe both changes within supply chains themselves as well as the evolution of processes, methods, and tools to manage them in this new "era". The growing popularity of collaborative platforms is highlighted by the rise of TradeCard's supply chain collaboration platform, which connects multiple buyers and suppliers with financial institutions, enabling them to conduct automated supply-chain finance transactions.

Web 2.0 is a trend in the use of the World Wide Web that is meant to increase creativity, information sharing, and collaboration among users. At its core, the common attribute of Web 2.0 is to help navigate the vast information available on the Web in order to find what is being bought. It is the notion of a usable pathway. SCM 2.0 replicates this notion in supply chain operations. It is the pathway to SCM results, a combination of processes, methodologies, tools, and delivery options to guide companies to their results quickly as the complexity and speed of the supply chain increase due to global competition; rapid price fluctuations; changing oil prices; short product life cycles; expanded specialization; near-, far-, and off-shoring; and talent scarcity.

SCM 2.0 leverages solutions designed to rapidly deliver results with the agility to quickly manage future change for continuous flexibility, value, and success. This is delivered through competency networks composed of best-of-breed supply chain expertise to understand which elements, both operationally and organizationally, deliver results, as well as through intimate understanding of how to manage these elements to achieve the desired results. The solutions are delivered in a variety of options, such as no-touch via business process outsourcing, mid-touch via managed services and software as a service (SaaS), or high-touch in the traditional software deployment model.

Business Process Integration

Successful SCM requires a change from managing individual functions to integrating activities into key supply chain processes. In an example scenario, a purchasing department places orders as its requirements become known. The marketing department, responding to customer demand, communicates with several distributors and retailers as it attempts to determine ways to satisfy this demand. Information shared between supply chain partners can only be fully leveraged through process integration.

Supply chain business process integration involves collaborative work between buyers and suppliers, joint product development, common systems, and shared information. According to Lambert and Cooper (2000), operating an integrated supply chain requires a continuous information flow. However, in many companies, management has concluded that optimizing product flows cannot be accomplished without implementing a process approach. The key supply chain processes stated by Lambert (2004) are:

- Customer relationship management
- Customer service management
- Demand management style
- Order fulfillment
- Manufacturing flow management
- Supplier relationship management
- Product development and commercialization
- Returns management

Much has been written about demand management. Best-in-class companies have similar characteristics, which include the following:

- Internal and external collaboration
- Initiatives to reduce lead time
- Tighter feedback from customer and market demand
- Customer-level forecasting

One could suggest other critical supply business processes that combine these processes stated by Lambert, such as:

Customer service management process

> Customer relationship management concerns the relationship between an organization and its customers. Customer service is the source of customer information. It also provides the customer with real-time information on scheduling and product availability through interfaces with the company's production and distribution operations. Successful organizations use the following steps to build customer relationships:
>
> - determine mutually satisfying goals for organization and customers
> - establish and maintain customer rapport
> - induce positive feelings in the organization and the customers

Procurement process

> Strategic plans are drawn up with suppliers to support the manufacturing flow management process and the development of new products. In firms whose operations extend globally, sourcing may be managed on a global basis. The

desired outcome is a relationship where both parties benefit and a reduction in the time required for the product's design and development. The purchasing function may also develop rapid communication systems, such as electronic data interchange (EDI) and Internet linkage, to convey possible requirements more rapidly. Activities related to obtaining products and materials from outside suppliers involve resource planning, supply sourcing, negotiation, order placement, inbound transportation, storage, handling, and quality assurance, many of which include the responsibility to coordinate with suppliers on matters of scheduling, supply continuity (inventory), hedging, and research into new sources or programs. Procurement has recently been recognized as a core source of value, driven largely by the increasing trends to outsource products and services, and the changes in the global ecosystem requiring stronger relationships between buyers and sellers.

Product development and commercialization

Here, customers and suppliers must be integrated into the product development process in order to reduce the time to market. As product life cycles shorten, the appropriate products must be developed and successfully launched with ever-shorter time schedules in order for firms to remain competitive. According to Lambert and Cooper (2000), managers of the product development and commercialization process must:

1. coordinate with customer relationship management to identify customer-articulated needs;
2. select materials and suppliers in conjunction with procurement; and
3. develop production technology in manufacturing flow to manufacture and integrate into the best supply chain flow for the given combination of product and markets.

Integration of suppliers into the new product development process was shown to have a major impact on product target cost, quality, delivery, and market share. Tapping into suppliers as a source of innovation requires an extensive process characterized by development of technology sharing, but also involves managing intellectual property issues.

Manufacturing flow management process

The manufacturing process produces and supplies products to the distribution channels based on past forecasts. Manufacturing processes must be flexible in order to respond to market changes and must accommodate mass customization. Orders are processes operating on a just-in-time (JIT) basis in minimum lot sizes. Changes in the manufacturing flow process lead to shorter cycle times, meaning improved responsiveness and efficiency in meeting customer demand. This process manages activities related to planning, scheduling, and supporting manufacturing operations, such as work-in-process storage, handling, transportation, and time phasing of components, inventory at manufacturing sites, and maximum flexibility in the coordination of geographical and final assemblies postponement of physical distribution operations.

Physical distribution

This concerns the movement of a finished product or service to customers. In physical distribution, the customer is the final destination of a marketing channel, and the availability of the product or service is a vital part of each channel participant's marketing effort. It is also through the physical distribution process that the time and space of customer service become an integral part of marketing. Thus it links a marketing channel with its customers (i.e., it links manufacturers, wholesalers, and retailers).

Outsourcing/partnerships

This includes not just the outsourcing of the procurement of materials and components, but also the outsourcing of services that traditionally have been provided in-house. The logic of this trend is that the company will increasingly focus on those activities in the value chain in which it has a distinctive advantage and outsource everything else. This movement has been particularly evident in logistics, where the provision of transport, storage, and inventory control is increasingly subcontracted to specialists or logistics partners. Also, managing and controlling this network of partners and suppliers requires a blend of central and local involvement: strategic decisions are taken centrally, while the monitoring and control of supplier performance and day-to-day liaison with logistics partners are best managed locally.

Performance measurement

Experts found a strong relationship from the largest arcs of supplier and customer integration to market share and profitability. Taking advantage of supplier capabilities and emphasizing a long-term supply chain perspective in customer relationships can both be correlated with a firm's performance. As logistics competency becomes a critical factor in creating and maintaining competitive advantage, measuring logistics performance becomes increasingly important, because the difference between profitable and unprofitable operations becomes narrower. A.T. Kearney Consultants (1985) noted that firms engaging in comprehensive performance measurement realized improvements in overall productivity. According to experts, internal measures are generally collected and analyzed by the firm, including cost, customer service, productivity, asset measurement, and quality. External performance is measured through customer perception measures and "best practice" benchmarking.

Warehousing management

> To reduce a company's cost and expenses, warehousing management is concerned with storage, reducing manpower cost, dispatching authority with on time delivery, loading & unloading facilities with proper area, inventory management system etc.

Workflow management

> Integrating suppliers and customers tightly into a workflow (or business process) and thereby achieving an efficient and effective supply chain is a key goal of workflow management.

Theories

There are gaps in the literature on supply chain management studies at present (2015): there is no theoretical support for explaining the existence or the boundaries of supply chain management. A few authors, such as Halldorsson et al. (2003), Ketchen and Hult (2006), and Lavassani et al. (2009), have tried to provide theoretical foundations for different areas related to supply chain by employing organizational theories, which may include the following:

- Resource-based view (RBV)
- Transaction cost analysis (TCA)
- Knowledge-based view (KBV)
- Strategic choice theory (SCT)
- Agency theory (AT)
- Channel coordination
- Institutional theory (InT)
- Systems theory (ST)
- Network perspective (NP)
- Materials logistics management (MLM)
- Just-in-time (JIT)
- Material requirements planning (MRP)
- Theory of constraints (TOC)
- Total quality management (TQM)

- Agile manufacturing
- Time-based competition (TBC)
- Quick response manufacturing (QRM)
- Customer relationship management (CRM)
- Requirements chain management (RCM)
- Dynamic Capabilities Theory
- Dynamic Management Theory
- Available-to-promise (ATP)
- Supply Chain Roadmap

However, the unit of analysis of most of these theories is not the supply chain but rather another system, such as the firm or the supplier-buyer relationship. Among the few exceptions is the relational view, which outlines a theory for considering dyads and networks of firms as a key unit of analysis for explaining superior individual firm performance (Dyer and Singh, 1998).

Supply Chain

In the study of supply chain management, the concept of centroids has become an important economic consideration. A centroid is a location that has a high proportion of a country's population and a high proportion of its manufacturing, generally within 500 mi (805 km). In the US, two major supply chain centroids have been defined, one near Dayton, Ohio, and a second near Riverside, California.

The centroid near Dayton is particularly important because it is closest to the population center of the US and Canada. Dayton is within 500 miles of 60% of the US population and manufacturing capacity, as well as 60% of Canada's population. The region includes the interchange between I-70 and I-75, one of the busiest in the nation, with 154,000 vehicles passing through per day, 30–35% of which are trucks hauling goods. In addition, the I-75 corridor is home to the busiest north-south rail route east of the Mississippi River.

Tax Efficient Supply Chain Management

Tax efficient supply chain management is a business model that considers the effect of tax in the design and implementation of supply chain management. As the consequence of globalization, cross-national businesses pay different tax rates in different countries. Due to these differences, they may legally optimize their supply chain and increase profits based on tax efficiency.

Sustainability and Social Responsibility in Supply Chains

Supply chain sustainability is a business issue affecting an organization's supply chain or logistics network, and is frequently quantified by comparison with SECH ratings, which uses a triple bottom line incorporating economic, social, and environmental aspects. SECH ratings are defined as social, ethical, cultural, and health' footprints. Consumers have become more aware of the environmental impact of their purchases and companies' SECH ratings and, along with non-governmental organizations (NGOs), are setting the agenda for transitions to organically grown foods, anti-sweatshop labor codes, and locally produced goods that support independent and small businesses. Because supply chains may account for over 75% of a company's carbon footprint, many organizations are exploring ways to reduce this and thus improve their SECH rating.

For example, in July 2009, Wal-Mart announced its intentions to create a global sustainability index that would rate products according to the environmental and social impacts of their manufacturing and distribution. The index is intended to create environmental accountability in Wal-Mart's supply chain and to provide motivation and infrastructure for other retail companies to do the same.

It has been reported that companies are increasingly taking environmental performance into account when selecting suppliers. A 2011 survey by the Carbon Trust found that 50% of multinationals expect to select their suppliers based upon carbon performance in the future and 29% of suppliers could lose their places on 'green supply chains' if they do not have adequate performance records on carbon.

The US Dodd–Frank Wall Street Reform and Consumer Protection Act, signed into law by President Obama in July 2010, contained a supply chain sustainability provision in the form of the Conflict Minerals law. This law requires SEC-regulated companies to conduct third party audits of their supply chains in order to determine whether any tin, tantalum, tungsten, or gold (together referred to as *conflict minerals*) is mined or sourced from the Democratic Republic of the Congo, and create a report (available to the general public and SEC) detailing the due diligence efforts taken and the results of the audit. The chain of suppliers and vendors to these reporting companies will be expected to provide appropriate supporting information.

Incidents like the 2013 Savar building collapse with more than 1,100 victims have led to widespread discussions about corporate social responsibility across global supply chains. Wieland and Handfield (2013) suggest that companies need to audit products and suppliers and that supplier auditing needs to go beyond direct relationships with first-tier suppliers. They also demonstrate that visibility needs to be improved if supply cannot be directly controlled and that smart and electronic technologies play a key role to improve visibility. Finally, they highlight that collaboration with local partners, across the industry and with universities is crucial to successfully managing social responsibility in supply chains.

Components

Management Components

SCM components are the third element of the four-square circulation framework. The level of integration and management of a business process link is a function of the number and level of components added to the link (Ellram and Cooper, 1990; Houlihan, 1985). Consequently, adding more management components or increasing the level of each component can increase the level of integration of the business process link.

Literature on business process re-engineering buyer-supplier relationships, and SCM suggests various possible components that should receive managerial attention when managing supply relationships. Lambert and Cooper (2000) identified the following components:

- Planning and control
- Work structure
- Organization structure
- Product flow facility structure
- Information flow facility structure
- Management methods
- Power and leadership structure
- Risk and reward structure
- Culture and attitude

However, a more careful examination of the existing literature leads to a more comprehensive understanding of what should be the key critical supply chain components, or "branches" of the previously identified supply chain business processes—that is, what kind of relationship the components may have that are related to suppliers and customers. Bowersox and Closs (1996) state that the emphasis on cooperation represents the synergism leading to the highest level of joint achievement. A primary-level channel participant is a business that is willing to participate in responsibility for inventory ownership or assume other financial risks, thus including primary level components (Bowersox and Closs, 1996). A secondary-level participant (specialized) is a business that participates in channel relationships by performing essential services for primary participants, including secondary level components, which support primary participants. Third-level channel participants and components that support primary-level channel participants and are the fundamental branches of secondary-level components may also be included.

Consequently, Lambert and Cooper's framework of supply chain components does not lead to any conclusion about what are the primary- or secondary-level (specialized) supply chain components —that is, which supply chain components should be viewed as primary or secondary, how these components should be structured in order to achieve a more comprehensive supply chain structure, and how to examine the supply chain as an integrative one.

Reverse Supply Chain

Reverse logistics is the process of managing the return of goods. It is also referred to as "aftermarket customer services". Any time money is taken from a company's warranty reserve or service logistics budget, one can speak of a reverse logistics operation. Reverse logistics is also the process of managing the return of goods from store, which the returned goods are sent back to warehouse and after that either warehouse scrap the goods or send them back to supplier for replacement depending on the warranty of the merchandise.

Systems and Value

Supply chain systems configure value for those that organize the networks. Value is the additional revenue over and above the costs of building the network. Co-creating value and sharing the benefits appropriately to encourage effective participation is a key challenge for any supply system. Tony Hines defines value as follows: "Ultimately it is the customer who pays the price for service delivered that confirms value and not the producer who simply adds cost until that point".

Global Applications

Global supply chains pose challenges regarding both quantity and value. Supply and value chain trends include:

- Globalization
- Increased cross-border sourcing
- Collaboration for parts of value chain with low-cost providers
- Shared service centers for logistical and administrative functions
- Increasingly global operations, which require increasingly global coordination and planning to achieve global optimums
- Complex problems involve also midsized companies to an increasing degree

These trends have many benefits for manufacturers because they make possible larger lot sizes, lower taxes, and better environments (e.g., culture, infrastructure, special tax zones, or sophisticated OEM) for their products. There are many additional challenges when the scope of supply chains is global. This is because with a supply chain of a larger scope, the lead time is much longer, and because there are more issues involved, such as multiple currencies, policies, and laws. The consequent problems include different currencies and valuations in different countries, different tax laws, different trading protocols, and lack of transparency of cost and profit.

Supply Chain Consulting

Supply-chain consulting is the providing of expert knowledge in order to assess the productivity of a supply-chain and, ideally, to enhance the productivity.

Supply chain Consulting is a service involved in transfer of knowledge on how to exploit existing assets through improved coordination and can hence be a source of competitive advantage; Hereby the role of the consultant is to help management by adding value to the whole process through the various sectors from the ordering of the raw materials to the final product.

On this regard, firms either build internal teams of consultants to tackle the issue or use external ones, (companies choose between these two approaches taking into consideration various factors).

The use of external consultants is a common practice among companies. The whole consulting process generally involves the analysis of the entire supply-chain process, including the countermeasures or correctives to take to achieve a better overall performance.

Certification

Skills and Competencies

Supply chain professionals need to have knowledge of managing supply chain functions such as transportation, warehousing, inventory management, and production planning. In the past, supply chain professionals emphasized logistics skills, such as knowledge of shipping routes, familiarity with warehousing equipment and distribution center locations and footprints, and a solid grasp of freight rates and fuel costs. More recently, supply chain management extends to logistical support across firms and management of global supply chains. Supply chain professionals need to have an understanding of business continuity basics and strategies.

Roles and Responsibilities

Supply chain professionals play major roles in the design and management of supply chains. In the design of supply chains, they help determine whether a product or service is provided by the firm itself (insourcing) or by another firm elsewhere (outsourcing). In the management of supply chains, supply chain professionals coordinate production among multiple providers, ensuring that production and transport of goods happen with minimal quality control or inventory problems. One goal of a well-designed and maintained supply chain for a product is to successfully build the product at minimal cost. Such a supply chain could be considered a competitive advantage for a firm.

Beyond design and maintenance of a supply chain itself, supply chain professionals participate in aspects of business that have a bearing on supply chains, such as sales forecasting, quality management, strategy development, customer service, and systems analysis. Production of a good may evolve over time, rendering an existing supply chain design obsolete. Supply chain professionals need to be aware of changes in production and business climate that affect supply chains and create alternative supply chains as the need arises. Individuals working in supply chain management can attain a professional certification by passing an exam developed by a third party certification organizations. The purpose of certification is to guarantee a certain level of expertise in the field.

Education

The knowledge needed to pass a certification exam may be gained from several sources. Some knowledge may come from college courses, but most of it is acquired from a mix of on-the-job learning experiences, attending industry events, learning best practices with their peers, and reading books and articles in the field. Certification organizations may provide certification workshops tailored to their exams. There are also free websites that provide a significant amount of educational articles, as well as blogs that are internationally recognized which provide good sources of news and updates.

Industrial Metabolism

Industrial metabolism is a concept to describe the material and energy turnover of industrial systems. It was proposed by Robert Ayres in analogy to the biological metabolism as "the whole integrated collection of physical processes that convert raw materials and energy, plus labour, into finished products and wastes" In analogy to the biological concept of metabolism, which is used to describe the whole of chemical reactions in, for example, a cell to maintain its functions and reproduce itself, the concept of industrial metabolism describes the chemical reactions, transport processes, and manufacturing activities in industry. Industrial metabolism presupposes a connection between different industrial activities by seeing them as part of a larger system, such as a material cycle or the supply chain of a commodity. System scientists, for example in industrial ecology, use the concept as paradigm to study the flow of materials or energy through the industrial system in order to better understand supply chains, the sources and causes of emissions, and the linkages between the industrial and the wider socio-technological system.

Industrial metabolism is a subsystem of the anthropogenic or socioeconomic metabolism, which also comprises non-industrial human activities in households or the public sector

Reuse

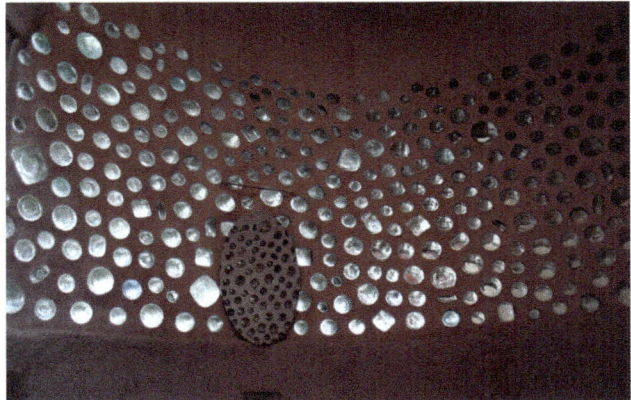

An interior bathroom wall that incorporates repurposed clear glass bottles into a bottle wall

Reuse is the action or practice of using something again, whether for its original purpose (conventional reuse) or to fulfil a different function (creative reuse or repurposing). It should be distinguished from recycling, which is the breaking down of used items to make raw materials for the manufacture of new products. Reuse – by taking, but not reprocessing, previously used items – helps save time, money, energy, and resources. In broader economic terms, it can make quality products available to people and organizations with limited means, while generating jobs and business activity that contribute to the economy.

Historically, financial motivation was one of the main drivers of reuse. In the developing world this driver can lead to very high levels of reuse, however rising wages and consequent consumer demand for the convenience of disposable products has made the reuse of low value items such as packaging uneconomic in richer countries, leading to the demise of many reuse programs. Current environmental awareness is gradually changing attitudes and regulations, such as the new packaging regulations, are gradually beginning to reverse the situation.

One example of conventional reuse is the doorstep delivery of milk in refillable bottles; other examples include the retreading of tires and the use of returnable/reusable plastic boxes, shipping containers, instead of single-use corrugated fiberboard boxes.

Advantages and Disadvantages

Ants, being social insects, have been able to reuse rail tracks abandoned by humans for their own transportation. (Kadina, South Australia)

Reuse has certain potential advantages:

- Energy and raw materials savings as replacing many single use products with one reusable one reduces the number that need to be manufactured.

- Reduced disposal needs and costs.

- Refurbishment can bring sophisticated, sustainable, well paid jobs to underdeveloped economies.

- Cost savings for business and consumers as a reusable product is often cheaper than the many single use products it replaces.

- Some older items were better handcrafted and appreciate in value.

Disadvantages are also apparent:

- Reuse often requires cleaning or transport, which have environmental costs.

- Some items, such as freon appliances, infant auto seats, older tube TVs and secondhand automobiles could be hazardous or less energy efficient as they continue to be used.

- Reusable products need to be more durable than single-use products, and hence require more material per item. This is particularly significant if only a small proportion of the reusable products are in fact reused.

- Sorting and preparing items for reuse takes time, which is inconvenient for consumers and costs money for businesses.

- Special skills are required to tweak the functional throughput of items when devoting them to new uses outside of their original purpose.

- Knowing the standards that legacy products conform to is required for knowing what adapters to buy for newer products to be compatible with them, even though the cost of adapters for such applications is a minor disadvantage.

- Being a rather minor disadvantage, metal that is repurposed later on can sometimes contain rust, seeing as it sometimes ages before reuse.

Examples

Reuse Centers and Virtual Exchange

A salvaged window from the deconstruction of an old house turned home decor with paint and stencils. Source: Habitat for Humanity Mt. Angel ReStore

These services facilitate the transaction and redistribution of unwanted, yet perfectly usable, materials and equipment from one entity to another. The entities that benefit from either side of this service (as donors, sellers, recipients, or buyers) can be businesses, nonprofits, schools, community groups, and individuals. Some maintain a physical space (a reuse center), and others act as a matching service (a virtual exchange). Reuse centers generally maintain both warehouses and trucks. They take possession of the donated materials and make them available for redistribution or sale. Virtual exchanges do not have physical space or trucks, but instead allow users to post listings of materials available and wanted (for free or at low cost) on an online materials exchange website. Staff will help facilitate the exchange of these materials without ever taking possession of the materials.

United States

- Goodwill Industries
- Salvation Army
- Second Harvest Food Bank
- Habitat for Humanity ReStores.

Virtual exchangees include:

- donateNYC (New York)
  - CalMax (California)
- Alchemy Goods
- Materials for the Arts (Queens, New York)
- STAY VOCAL (Norwell, MA)

Consumer resources exist for exchanging usable materials, such as freecycling sites which are often grassroots and entirely nonprofit movements of people who are giving (and getting) stuff for free in their own towns. It's all about reuse and keeping good stuff out of landfills. Membership is normally free. In, addition, there are directory-based resources such as RecycleChicken.com which point consumers to local and national locations for reuse and repurposing of materials not normally accepted in recycling programs.

- Teleplan Camera Repair has introduced a free camera recycling program through the reuse of cameras.

Australia

- Woolloongabba, Queensland
  - Reverse Garbage Queensland
- Marrickville, New South Wales:
  - Reverse Garbage is the largest reuse centre in the country, diverting more than 12,000 cubic metres of resources from landfills each year.
  - The Bower Reuse and Repair Centre diverts more than 7,500 cubic metres of 'waste' from landfills a year in a building entirely made of salvaged materials.
- Perth, Western Australia:
  - REmida WA

With technological innovations, new applications and shorter product lifetimes there is an ever increasing quantity of waste electrical and electronic equipment. The environmental pressures associated with this are well documented and include material and energy losses and an increase in air, water and land pollution from waste treatment methods. The incorrect disposal and treatment of electrical and electronic equipment also poses threats to human health, particularly when involving illegal exports.

Addressing Issues of Repair, Reuse and Recycling

One way to address this is to increase product longevity; either by extending a product's first life or addressing issues of repair, reuse and recycling. Reusing products, and therefore extending the use of that item beyond the point where it is discarded by its first user is preferable to recycling or disposal, as this is the least energy intensive solution, although it is often overlooked.

The EU Circular Economy Package recognises the importance of extending product lifetimes and includes repair and reuse of products in its action plan to ensure products reach their optimum lifespan. If targets for reducing greenhouse gas emissions are to be reached, then reuse needs to be included as part of a whole life cycle approach.

A strong second hand market-place exists, with charity shops on most high streets, car boot sales and online auction sites maintaining popularity and regular TV shows featuring both buying and selling at auction.

Inadequate repair infrastructure

However, the reuse of electrical and electronic equipment products remains low due to inadequate infrastructure, lack of repair networks and products. Local authority collection systems are better suited for handling waste than handling goods and preserving reuse potential. Retailer delivery staff are trained to handle goods carefully.

So, do we need a radical rethink on how we move unwanted, still useable items to the second-hand market place? Is there a case for changing our approach to producer responsibility and insisting that producers finance collection for reuse, and additionally, drive consumer choices for reuse, repair and remanufacture; whilst addressing the costs of recycling and disposal?

There are opportunities for producers, waste management companies and local authorities to make both repair and reuse habitual, whilst these require changes to householder behaviour change through raising awareness, they also require investment in infrastructure and logistical operations. Is it time to insist that more products are designed to have longer lifetimes? That they can be disassembled, repaired and reused before being recycled?

Business models providing opportunities

This would not necessarily be a poor strategy for businesses, there are business models that provide opportunities to retain ownership of valuable products and components through leasing, servicing, repair and re-sale.

While it is choices made by consumers that will ultimately determine the success of such ventures, there is huge potential for the reuse of goods and materials to deliver social and economic and environmental benefits. The EU Circular Economy Package, the Scottish Circular Economy Strategy and the national reuse target set by the Spanish Government are examples of governments recognising that second-hand goods should be a good value mainstream option and are working towards making reuse easier for consumers.

In environmental terms, reuse ought to be more common than recycling and energy recovery, with both the financial and environmental costs of simple refurbishment of some products being a fraction of original manufacturing costs. If we are going to be serious about living in a Circular Economy we need to recognise the value of our waste and ensure resources are kept in the economy for longer, slow down the use of valuable raw materials and ensure that products are reused and materials are recycled rather than landfilled.

Remanufacturing

The most involved reuse organizations are "repair and overhaul" industries which take valuable parts, such as engine blocks, office furniture, toner cartridges, single-use cameras, aircraft hulls, and cathode ray tubes (CRTs) and refurbish them in a factory environment in order to meet the same/similar specifications as new products. Xerox (copy machines), and Cummins Engine are examples of refurbishing factories in the USA. Rolls Royce has a very large aircraft remanufacturing factory in Singapore; Caterpillar recently announced the opening of a tractor refurbishing plant in China. Some factories operate in competition with the original equipment manufacturer (OEM). When the refurbished item is resold under a new label (used monitor CRTs made into TVs, or cameras resold under a new label) this has been found legal by most courts.

When the item is resold under the same OEM name, it is informally considered a "gray market" item - if it is sold as used, it's legal, if it's represented as an OEM product eligible for rebates and warranties, it is considered "counterfeit" or "black market". The automobile parts industry in the USA is governed by laws on the disclosure of "used" parts and, in some states, mattresses which have been used are required to be sanitized or destroyed. Whether these laws are in place to protect consumers from black market items, or to protect manufacturers ("hindsight obsolescence"), is often an area of intense debate. Fuji Photo Film Co. v. Jazz Photo Corp. is a recent example of the war between patent holders and refurbishing factories. To quote the 2003 District Court of New Jersey:*"Thus, the key issue in the dispute between Fuji and Jazz is whether the cameras sold by Jazz are "refurbished" in such a way that they can be considered to have been permissibly "repaired" or impermissibly "reconstructed."*

Package Deposit Programs

Reusable glass bottles collected in Bishkek, Kyrgyzstan. Deposit values (0.5-2 Kyrgyz som) are posted next to the sample bottles on the rack

Deposit programs offer customers a financial incentive to return packaging for reuse. Although no longer common, international experience is showing that they can still be an effective way to encourage packaging reuse. However, financial incentive, unless great, may be less of an incentive than convenience: statistics show that, on average, a milk bottle is returned 12 times, whereas a lemonade bottle with a 15p deposit is returned, on average, only 3 times.

Refillable bottles are used extensively in many European countries; for example in Denmark, 98% of bottles are refillable, and 98% of those are returned by consumers. These systems are typically supported by deposit laws and other regulations.

Sainsbury Ltd have operated a plastic carrier bag cash refund scheme in 1991 - "the penny back scheme". The scheme is reported to save 970 tonnes of plastic per annum. The scheme has now been extended to a penny back on a voucher which can be contributed to schools registered on the scheme; it estimates this will raise the savings in plastic to 2500 tonnes per annum.

The 600 ml brown bottle is the "standard beer reused bottle" in Brazil.

In some developing nations like India and Pakistan, the cost of new bottles often forces manufacturers to collect and refill old glass bottles for selling cola and other drinks. India and Pakistan also have a way of reusing old newspapers: "Kabadiwalas" buy these from the readers for scrap value and reuse them as packaging or recycle them. Scrap intermediaries help consumer dispose of other materials including metals and plastics.

Closed-loop Programs

These apply primarily to items of packaging, for example, where a company is involved in the regular transportation of goods from a central manufacturing facility to warehouses or warehouses to retail outlets. In these cases there is considerable benefit to using reusable "transport packaging" such as plastic crates or pallets.

The benefits of closed-loop reuse are primarily due to low additional transport costs being involved, the empty lorry returning with the empty crates. There have been some recent attempts to get the public to join in on closed loop reuse schemes where shoppers use reusable plastic baskets in place of carrier bags for transporting their goods home from the supermarket; these baskets fit on specially designed trolleys making shopping supposedly easier.

Refilling Programs

There have been some market-led initiatives to encourage packaging reuse by companies introducing refill packs of certain commodities (mainly soap powders and cleaning fluids), the contents being transferred before use into a reusable package kept by the customer, with the savings in packaging being passed onto the customer by lower shelf prices. The refill pack itself is not reused, but being a minimal package for carrying the product home, it requires less material than one with the durability and features (reclosable top, convenient shape, etc.) required for easy use of the product, while avoiding the transport cost and emissions of returning the reusable package to the factory.

Regifting

Some items, such as clothes and children's toys, often become unwanted before they wear out due to changes in their owner's needs or preferences; these can be reused by selling or giving them to new owners. Regiving can take place informally between family, friends, or neighbours, through environmental freecycling organisations or through anti-poverty charities such as the Red Cross, United Way, Salvation Army, and Goodwill which give these items to those who could not afford them new. Other organizations such as iLoveSchools have websites where both new and used goods can be offered to any of America's school teachers so their life can be extended and help schoolchildren. The average American, for example, throws away 67.9 pounds of used clothing and rags. With the U.S. population at approximately 296 million people, that translates into 20 billion pounds of used clothing and textiles that are tossed into the landfills each year. This has partly motivated movements such as The Compact, whose members promise not to buy anything new for a year, and rely on reusing items that otherwise would be thrown away. Reuse not only reduces landfill inline with the waste minimization program but can help raise money for a good cause.

Printer Cartridges and Toners

Printer ink cartridges can be reused. They are sorted by brand and model, to be refilled or resold back to the manufacturers. The companies then refill the ink reservoir to resell to consumers. Toner cartridges are recycled the same way as ink cartridges, using toner instead of ink. This method is highly efficient as there is no energy spent on melting and recreating the cartridges.

Repurposing

Plastic bottles (with LED lights) repurposed as a chandelier during Ramadan in the Muslim Quarter, Jerusalem

Repurposing is to use a tool for use as another tool, usually for a purpose unintended by the original tool-maker. Typically, repurposing is done using items usually considered to be junk or garbage. A good example of this would be the Earthship style of house, that uses tires as insulating walls and bottles as glass walls. Reuse is not limited to repeated uses for the same purpose. Examples of repurposing include using tires as boat fenders and steel drums or plastic drums as feeding troughs and/or composting bins.

Waste Exchanges

A waste exchange, or virtual exchange, facilitates the use of a waste product from one process as a raw material for another. As with new life reuse of finished items, this avoids the environmental costs of disposing of the waste and obtaining new raw material, and may still be possible if the nature of the process makes avoiding production of the waste or recycling it back into the original process impossible.

This sort of scheme needs to have a far broader base than is currently the case, it requires organization and the setting up of waste brokerages where lists of currently available wastes are and the quantities available. One of the problems is once a demand for a waste is known or shown then the material is no longer a "waste" but a sellable commodity which often prices itself out of the market, c.f waste cement kiln dust and N-viro (lime conditioned sewage sludge fertilizer). In the former East Germany, organic household waste was collected and used as fodder for pigs. This integrated system was made possible by the state's control of agriculture; the complexities of continuing it in a market economy after German reunification meant the system had to be discontinued.

Reuse of Waste Water and Excreta in Agriculture

The nutrients, i.e. nitrogen, phosphorus, potassium and micronutrients, and organic matter contained in wastewater, excreta (urine and feces) and greywater have traditionally been reused in agriculture in many countries and are still being reused in agriculture to this day - unfortunately often in an unregulated and unsafe manner for example in many developing countries (e.g. Mexico, India, Bangladesh, Ghana). The WHO Guidelines from 2006 have set up a framework how this reuse can be done safely by following a multiple barrier approach. Work by the International Water Management Institute has led to a better understanding on how such wastewater reuse can be safely implemented in practice, for which they won the Stockholm Water Prize in 2012. Reuse of sanitised excreta in agriculture has also been called a "closing the loop" approach for sanitation and agriculture and is central to the ecological sanitation approach.

Measuring the Impact of Reuse, Reuse Metrics

Determining the balance of how the several effects of reuse interact is often best accomplished by a formal life cycle assessment. For example, research has shown that reusing a product can reduce CO_2 emissions and carbon footprint by more than 50% relative to the complete product life cycle. A relatively unknown effective way to reduce CO_2 emissions and carbon footprint is reusing products. Often the relative carbon footprint of manufacturing and the supply chain is unknown. A scientific methodology has been developed to calculate how much CO_2 emissions are reduced when buying used or second hand hardware versus new hardware, the so-called durability greener network calculator.

There are many ways of measuring the positive environmental, economic and social impact data. These include:

- # of tons diverted from the landfill
- $ avoided disposal costs
- $ avoided purchase costs
- $ value of materials donated
- $ revenues earned
- # of jobs created or retained
- # of families/individuals/organizations assisted

Internalized Environmental Costs

A Pigovian tax is an environmental tax: a charge on items that reflects the environmental costs of their manufacture and disposal. This makes the environmental benefit of using one reusable item instead of many disposable ones into a financial incentive. Such charges have been introduced in some countries.

Comparison to Recycling

A school being prepared for reuse as housing

Recycling differs from reuse in that it breaks down the item into raw materials which are then used to make new items, as opposed to reusing the intact item. As this extra processing requires energy, as a rule of thumb reuse is environmentally preferable to recycling ("reduce, reuse, recycle"), though recycling does have a significant part to play as it can often make use of items which are broken, worn out or otherwise unsuitable for reuse. However, as transport emissions are significant portion of the environmental impact of both reuse and recycling, in some cases recycling is the more prudent course as reuse can require long transport distances. A complex life cycle analysis may be required during a product's design phase to determine the efficacy of reuse, recycling, or neither, and produce accordingly.

Reuse of Information

Besides physical resources, information is often reused, notably program code for the software that drives computers and the Internet, but also the documentation that explains how to use every modern device. And it is proposed as a way to improve education by assembling a great library of shareable learning objects that can be reused in learning management systems.

Software reuse grew out of the standard subroutine libraries of the 1960s. It is the main principle of today's object-oriented programming. Instead of constantly reinventing software wheels, programming languages like C++, Java, Objective-C, and others are building vast collections of reusable software objects and components.

Reuse of information has a tremendous return on investment for organizations whose documentation is translated into many languages. Translation memory systems can store text that has already been translated into dozens of languages for retrieval and reuse.

Reuse of Older Software

Reuse of older software is popular among retrocomputing. Most of the time, emulators are used to run older software from other platforms, or other operating systems.

Sometimes, older operating systems such as DOS are reused for computing roles that don't demand lots of computing power. However, the widespread of availability of secondhand Windows XP computers at extreme low prices has largely supplanted immediate opportunities for using DOS on some repurposing applications, especially since something like USB isn't found on most pre-Windows XP computers.

Remanufacturing

Remanufacturing is the rebuilding of a product to specifications of the original manufactured product using a combination of reused, repaired and new parts. It requires the repair or replacement of worn out or obsolete components and modules. Parts subject to degradation affecting the performance or the expected life of the whole are replaced. Remanufacturing is a form of a product recovery process that differs from other recovery processes in its completeness: a remanufactured machine should match the same customer expectation as new machines.

In 1995, the United States Environmental Protection Agency (EPA) implemented the Comprehensive Procurement Guideline (CPG) program to promote waste reduction and resource conservation through the use of materials recovered from solid waste, and to ensure that the materials collected in recycling programs will be used again in the manufacture of new products. The EPA is required to designate products that are or can be made with recovered materials, and to recommend practices for buying these products. Once a product is designated, state and federal procuring agencies are required to purchase it with the highest recovered material content level practicable.

In 2004, the EPA published its third CPG update (CPG IV) which designated seven additional products and revised three existing product designations. One of the new product categories to be added was Rebuilt Vehicular Parts. The EPA defines rebuilt vehicular parts as "vehicle parts that have been re-manufactured, reusing parts in their original form. Rebuilt parts undergo an extensive re-manufacturing and testing process and must meet the same industry specifications for performance as new parts."

In the UK, a market potential of up to 5.6 billion GBP has been identified in remanufacturing, with the benefits said to be improvement to business margins, revenues and security of supply.

Other Forms of Product Recovery

1. Reuse implies that items are used by a second customer without prior repair operations or as originally designed.

2. Repair: the process of bringing damaged components back to a functional condition.

3. Refurbishing/Reconditioning is the process of restoring components to a functional and/or satisfactory state to the original specification, using methods such as resurfacing, repainting, etc.

4. Recycling is the process of taking a component material and processing it to make the same material or useful degraded material.

5. Cannibalization (parts)

Many formal definitions of remanufacturing exist in the literature, but the first published report on remanufacturing, by R. Lund (1984), describes remanufacturing as " an industrial process in which worn-out products are restored to like-new condition. Through a series of industrial processes in a factory environment, a discarded product is completely disassembled. Useable parts are cleaned, refurbished, and put into inventory. Then the product is reassembled from the old parts (and where necessary, new parts) to produce a unit fully equivalent and sometimes superior in performance and expected lifetime to the original new product".

Furthermore, the Automotive Parts Remanufacturers Association (APRA) realized that communication problems can arise when people from different countries with different language skills talk about remanufacturing. Certain terms can have different meanings as definitions between countries and individuals vary. In 2013, APRA was able to solve these communication problems by publishing a common translation list in many different languages in order to unite all those who deal with the automotive industry.

Range of Products Being Remanufactured

- Aerospace
- Air-conditioning units
- Bakery Equipment
- Carpet tiles
- Compressors
- Computer and telecommunication equipment.
- Defense equipment
- Electrical motors and apparatus
- Excavation equipment
- Fashion apparel and accessories
- Gaming Machines
- Industrial food processing equipment
- Machine tools
- Medical Equipment
- Musical Instruments
- Office furniture
- Office photocopiers (laser toner cartridges)
- Power bearings

- Pumps
- Robots
- Rolling stock (railway vehicles)
- Surgical Medical Tables
- Toner Cartridges
- Vehicular Parts
- Vending Machines

Different Types of Remanufacturing

There are three types of remanufacturing activities, each with different operational challenges.

1. *Remanufacturing without identity loss*; with this method, a current machine is built on yesterday's base, receiving all of the enhancements, expected life and warranty of a new machine. The physical structure (the chassis or frame) is inspected for soundness. The whole product is refurbished and critical modules are overhauled, upgraded or replaced. If there are defects in the original design, they are eliminated. This is the case for customized remanufacturing of machine tools, airplanes, computer mainframes, large medical equipment and other capital goods. Because of its uniqueness, this product recovery is characterized as a project.

2. *Remanufacturing by recoating of worn engine parts*; many engine parts, components are large and expensive and after a period of use become worn. An example of such a part is the engine block, in particular the cylinder engine bores, which must withstand explosions during piston firing. Instead of disposing of large engine blocks, remanufacturing has resulted in re-use of the parts by coating them with plasma transferred wire arc spraying (PTWA) Caterpillar known for manufacturing very large industrial trucks and machinery has started such remanufacturing programs of equipment parts using PTWA, resulting in a greener environment. Remanufacturing by recoating of parts is also very popular in the aircraft field, the geothermal pipe field and the automotive engine field.

3. *Repetitive remanufacturing without identity loss*; in this method, there is the additional challenge of scheduling the sequence of dependent processes and identifying the location of inventory buffers. There is a fine line between repetitive remanufacturing without loss of identity and product overhaul. Again, the critical difference is that remanufacturing is a complete process. The final output has a like-new appearance and is covered by a warranty comparable to that of a new product.

Remanufacturing with Loss of Original Product Identity

With this method, used goods are disassembled into pre-determined components and repaired to stock, ready to be reassembled into a remanufactured product. This is the case when remanufacturing automobile components, photocopiers, toner cartridges, furniture, ready-to-use cameras and personal computers. Once the product is disassembled and the parts are recovered, the process concludes with an operation not too different from original manufacturing. Disassembled parts are inventoried, just like purchased parts and made available for final assembly.

Remanufacturing with loss of original product identity encompasses some unique challenges in inventory management and disassembly sequence development. Some of the open questions relate to the commonality of parts in products of different generations, the uncertainty in the supply of used products, and their relationship with production planning. The National Center for Remanufacturing and Resource Recovery (NCR3) at Rochester Institute of Technology (NY) is researching remanufacturing processes including testing standards for remanufactured products.

Rebuilding

Rebuilding is an old name for remanufacturing. It is still widely used by automotive industry. For example, the Automotive Parts Remanufacturers Association (APRA), have the new term in their name, but to be safe on their own website use the combined term 'rebuild/remanufacture'.

The term 'rebuilding' is also often used by railway companies; a steam locomotive may be rebuilt with a new boiler or a diesel locomotive may be rebuilt with a new engine. This saves money (by re-using the frame, and some other components, which still have years of useful life) and allows the incorporation of improved technology. For example, a new diesel engine may have lower fuel consumption, reduced exhaust emissions and better reliability. Recent examples include British Rail Class 57 and British Rail Class 43.

Benefits of Engine Remanufacturing

1. Lower Cost - One of the biggest factors in choosing a remanufactured engine is cost. Remanufactured engines cost up to 50 percent less than a new Tier 4 Final engine, not including associated engineering costs, and offers even greater savings over the purchase of a new piece of equipment. Maintenance on some new Tier 4 Final engines can be costly as well. In addition to being more complex to service, additional costs for after treatment cleaning and DEF fluid can add up quickly.

2. Less Downtime - Opting to replace your existing engine with a new one typically requires significant engineering time that can render rental equipment out of

service for months. The integration and testing process also takes significantly longer with a new Tier 4 Final engine than with remanufactured engines.

3. Increased Equipment Resale Value - There are two main factors currently making it difficult to sell Tier 4 Final-powered equipment in lesser regulated countries: serviceability and fuel availability. The lack of established service programs for the new technology makes supporting the product difficult. This coupled with high product costs, have companies shying away from Tier 4 Final powered equipment. In addition, many lesser regulated countries don't have high availability for ultra-low sulphur diesel fuel, which is required by Tier 4 Final engines to operate effectively. Without it, the diesel particulate filter can clog rapidly and cause significant operation issues.

4. More Sustainable Option - Purchasing a remanufactured engine keeps an old engine core and many usable components out of a landfill, which can greatly reduce the impact on the environment. According to Perkins Pacific's Andy Machin, approximately 73 percent of the old engine can be salvaged during the remanufacturing process saving millions of pounds of waste out of landfills. New engine production requires all new materials, substantial amounts of energy for the production process and additional transportation costs. By utilizing recycled components, a remanufactured engine typically uses about 80 percent less energy than new engine production, making it a much greener option. For rental fleet owners, reman offers the opportunity to expand their sustainability practices while saving money.

Recycling

The three chasing arrows of the international recycling logo.
It is sometimes accompanied by the text "reduce, reuse, and recycle".

Recycling is the process of converting waste materials into new materials and objects. It is an alternative to "conventional" waste disposal that can save material and help lower greenhouse gas emissions (compared to plastic production, for example). Recycling can prevent the waste of potentially useful materials and reduce the consumption of fresh raw materials, thereby reducing: energy usage, air pollution (from incineration), and water pollution (from landfilling).

Recycling is a key component of modern waste reduction and is the third component of the "Reduce, Reuse, and Recycle" waste hierarchy.

There are some ISO standards related to recycling such as ISO 15270:2008 for plastics waste and ISO 14001:2004 for environmental management control of recycling practice.

Recyclable materials include many kinds of glass, paper, and cardboard, metal, plastic, tires, textiles, and electronics. The composting or other reuse of biodegradable waste—such as food or garden waste—is also considered recycling. Materials to be recycled are either brought to a collection centre or picked up from the curbside, then sorted, cleaned, and reprocessed into new materials destined for manufacturing.

In the strictest sense, recycling of a material would produce a fresh supply of the same material—for example, used office paper would be converted into new office paper or used polystyrene foam into new polystyrene. However, this is often difficult or too expensive (compared with producing the same product from raw materials or other sources), so "recycling" of many products or materials involves their *reuse* in producing different materials (for example, paperboard) instead. Another form of recycling is the salvage of certain materials from complex products, either due to their intrinsic value (such as lead from car batteries, or gold from circuit boards), or due to their hazardous nature (e.g., removal and reuse of mercury from thermometers and thermostats).

History

Recycling has been a common practice for most of human history, with recorded advocates as far back as Plato in 400 BC. During periods when resources were scarce and hard to come by, archaeological studies of ancient waste dumps show less household waste (such as ash, broken tools, and pottery)—implying more waste was being recycled in the absence of new material.

In pre-industrial times, there is evidence of scrap bronze and other metals being collected in Europe and melted down for perpetual reuse. Paper recycling was first recorded in 1031 when Japanese shops sold repulped paper. In Britain dust and ash from wood and coal fires was collected by "dustmen" and downcycled as a base material used in brick making. The main driver for these types of recycling was the economic advantage of obtaining recycled feedstock instead of acquiring virgin material, as well as a lack of public waste removal in ever more densely populated areas. In 1813, Benjamin Law developed the process of turning rags into "shoddy" and "mungo" wool in Batley, Yorkshire. This material combined recycled fibers with virgin wool. The West Yorkshire shoddy industry in towns such as Batley and Dewsbury lasted from the early 19th century to at least 1914.

Industrialization spurred demand for affordable materials; aside from rags, ferrous scrap metals were coveted as they were cheaper to acquire than virgin ore. Railroads both purchased and sold scrap metal in the 19th century, and the growing steel and automobile industries purchased scrap in the early 20th century. Many secondary goods were collected, processed and sold by peddlers who scoured dumps and city streets for discarded machinery, pots, pans, and other sources of metal. By World War I, thousands of such peddlers roamed the streets of American cities, taking advantage of market forces to recycle post-consumer materials back into industrial production.

An American poster from World War II

British poster from World War II

Beverage bottles were recycled with a refundable deposit at some drink manufacturers in Great Britain and Ireland around 1800, notably Schweppes. An official recycling system with refundable deposits was established in Sweden for bottles in 1884 and aluminum beverage cans in 1982; the law led to a recycling rate for beverage containers of 84–99 percent depending on type, and a glass bottle can be refilled over 20 times on average.

Wartime

New chemical industries created in the late 19th century both invented new materials (e.g. Bakelite (1907) and promised to transform valueless into valuable materials. Proverbially, you could not make a silk purse of a sow's ear—until the US firm Arhur D. Little published in 1921 "On the Making of Silk Purses from Sows' Ears", its research proving that when "chemistry puts on overalls and gets down to business . . .new values appear. New and better paths are opened to reach the goals desired."

Recycling was a highlight throughout World War II. During the war, financial constraints and significant material shortages due to war efforts made it necessary for countries to reuse goods and recycle materials. These resource shortages caused by the world wars, and other such world-changing occurrences, greatly encouraged recycling. The struggles of war claimed much of the material resources available, leaving little for the civilian population. It became necessary for most homes to recycle their waste, as recycling offered an extra source of materials allowing people to make the most of what was available to them. Recycling household materials meant more resources for war efforts and a better chance of victory. Massive government promotion campaigns were carried out in the home front during World War II in every country involved in the war, urging citizens to donate metals and conserve fiber, as a matter of patriotism.

Post-war

A considerable investment in recycling occurred in the 1970s, due to rising energy costs. Recycling aluminum uses only 5% of the energy required by virgin production; glass, paper and other metals have less dramatic but very significant energy savings when recycled feedstock is used.

Although consumer electronics such as the television have been popular since the 1920s, recycling of them was almost unheard of until early 1991. The first electronic waste recycling scheme was implemented in Switzerland, beginning with collection of old refrigerators but gradually expanding to cover all devices. After these schemes were set up, many countries did not have the capacity to deal with the sheer quantity of e-waste they generated or its hazardous nature. They began to export the problem to developing countries without enforced environmental legislation. This is cheaper, as recycling computer monitors in the United States costs 10 times more than in China. Demand in Asia for electronic waste began to grow when scrap yards found that they could extract valuable substances such as copper, silver, iron, silicon, nickel, and gold, during the recycling process. The 2000s saw a large increase in both the sale of electronic devices and their growth as a waste stream: in 2002, e-waste grew faster than any other type of waste in the EU. This caused investment in modern, automated facilities to cope with the influx of redundant appliances, especially after strict laws were implemented in 2003.

As of 2014, the European Union has about 50% of world share of the waste and recycling industries, with over 60,000 companies employing 500,000 persons, with a turnover of €24 billion. Countries have to reach recycling rates of at least 50%, while the lead countries are around 65% and the EU average is 39% as of 2013.

Legislation

Supply

For a recycling program to work, having a large, stable supply of recyclable material is crucial. Three legislative options have been used to create such a supply: mandatory recycling collection, container deposit legislation, and refuse bans. Mandatory collection laws set recycling targets for cities to aim for, usually in the form that a certain percentage of a material must be diverted from the city's waste stream by a target date. The city is then responsible for working to meet this target.

Container deposit legislation involves offering a refund for the return of certain containers, typically glass, plastic, and metal. When a product in such a container is purchased, a small surcharge is added to the price. This surcharge can be reclaimed by the consumer if the container is returned to a collection point. These programs have been very successful, often resulting in an 80 percent recycling rate. Despite such good results, the shift in collection costs from local government to industry and consumers has created strong opposition to the creation of such programs in some areas. A variation on this is where the manufacturer bears responsibility for the recycling of their goods. In the European Union, the WEEE Directive requires producers of consumer electronics to reimburse the recyclers' costs.

An alternative way to increase supply of recyclates is to ban the disposal of certain materials as waste, often including used oil, old batteries, tires, and garden waste. One aim of this method is to create a viable economy for proper disposal of banned products. Care must be taken that enough of these recycling services exist, or such bans simply lead to increased illegal dumping.

Government-mandated Demand

Legislation has also been used to increase and maintain a demand for recycled materials. Four methods of such legislation exist: minimum recycled content mandates, utilization rates, procurement policies, and recycled product labeling.

Both minimum recycled content mandates and utilization rates increase demand directly by forcing manufacturers to include recycling in their operations. Content mandates specify that a certain percentage of a new product must consist of recycled material. Utilization rates are a more flexible option: industries are permitted to meet the recycling targets at any point of their operation or even contract recycling out in exchange for tradeable credits. Opponents to both of these methods point to the large increase in reporting requirements they impose, and claim that they rob industry of necessary flexibility.

Governments have used their own purchasing power to increase recycling demand through what are called "procurement policies." These policies are either "set-asides," which reserve a certain amount of spending solely towards recycled products, or "price preference" programs which provide a larger budget when recycled items are purchased. Additional regulations can target specific cases: in the United States, for example, the Environmental Protection Agency mandates the purchase of oil, paper, tires and building insulation from recycled or re-refined sources whenever possible.

The final government regulation towards increased demand is recycled product labeling. When producers are required to label their packaging with amount of recycled material in the product (including the packaging), consumers are better able to make educated choices. Consumers with sufficient buying power can then choose more environmentally conscious options, prompt producers to increase the amount of recycled material in their products, and indirectly increase demand. Standardized recycling labeling can also have a positive effect on supply of recyclates if the labeling includes information on how and where the product can be recycled.

Recyclates

Glass recovered by crushing only one kind of beer bottle

Recyclate is a raw material that is sent to, and processed in a waste recycling plant or materials recovery facility which will be used to form new products. The material is collected in various methods and delivered to a facility where it undergoes re-manufacturing so that it can be used in the production of new materials or products. For example, plastic bottles that are collected can be re-used and made into plastic pellets, a new product.

Quality of Recyclate

The quality of recyclates is recognized as one of the principal challenges that needs to be addressed for the success of a long-term vision of a green economy and achieving zero waste. Recyclate quality is generally referring to how much of the raw material is made up of target material compared to the amount of non-target material and other non-recyclable material. Only target material is likely to be recycled, so a higher amount of non-target and non-recyclable material will reduce the quantity of recycling product. A high proportion of non-target and non-recyclable material can make it more difficult for re-processors to achieve "high-quality" recycling. If the recyclate is of poor quality, it is more likely to end up being down-cycled or, in more extreme cases, sent to other recovery options or landfilled. For example, to facilitate the re-manufacturing of clear glass products there are tight restrictions for colored glass going into the re-melt process.

The quality of recyclate not only supports high-quality recycling, but it can also deliver significant environmental benefits by reducing, reusing and keeping products out of landfills. High-quality recycling can help support growth in the economy by maximizing the economic value of the waste material collected. Higher income levels from the sale of quality recyclates can return value which can be significant to local governments, households, and businesses. Pursuing high-quality recycling can also provide consumer and business confidence in the waste and resource management sector and may encourage investment in that sector.

There are many actions along the recycling supply chain that can influence and affect the material quality of recyclate. It begins with the waste producers who place non-target and non-recyclable wastes in recycling collection. This can affect the quality of final recyclate streams or require further efforts to discard those materials at later stages in the recycling process. The different collection systems can result in different levels of contamination. Depending on which materials are collected together, extra effort is required to sort this material back into separate streams and can significantly reduce the quality of the final product. Transportation and the compaction of materials can make it more difficult to separate material back into separate waste streams. Sorting facilities are not one hundred per cent effective in separating materials, despite improvements in technology and quality recyclate which can see a loss in recyclate quality. The storage of materials outside where the product can become wet can cause problems for re-processors. Reprocessing facilities may require further sorting steps to further reduce the amount of non-target and non-recyclable material. Each action along the recycling path plays a part in the quality of recyclate.

Quality Recyclate Action Plan (Scotland)

The Recyclate Quality Action Plan of Scotland sets out a number of proposed actions that the Scottish Government would like to take forward in order to drive up the quality of the materials being collected for recycling and sorted at materials recovery facilities before being exported or sold on to the reprocessing market.

The plan's objectives are to:

- Drive up the quality of recyclate.

- Deliver greater transparency about the quality of recyclate.

- Provide help to those contracting with materials recycling facilities to identify what is required of them

- Ensure compliance with the Waste (Scotland) regulations 2012.

- Stimulate a household market for quality recyclate.

- Address and reduce issues surrounding the Waste Shipment Regulations.

The plan focuses on three key areas, with fourteen actions which were identified to increase the quality of materials collected, sorted and presented to the processing market in Scotland.

The three areas of focus are:

1. Collection systems and input contamination
2. Sorting facilities – material sampling and transparency
3. Material quality benchmarking and standards

Recycling Consumer Waste

Collection

A three-sided bin at a railway station in Germany, intended to separate paper *(left)* and plastic wrappings *(right)* from other waste *(back)*

A number of different systems have been implemented to collect recyclates from the general waste stream. These systems lie along the spectrum of trade-off between public convenience and government ease and expense. The three main categories of collection are "drop-off centers," "buy-back centers", and "curbside collection."

Curbside Collection

Curbside collection encompasses many subtly different systems, which differ mostly on where in the process the recyclates are sorted and cleaned. The main categories are mixed waste collection, commingled recyclables, and source separation. A waste collection vehicle generally picks up the waste.

A recycling truck collecting the contents of a recycling bin in Canberra, Australia

At one end of the spectrum is mixed waste collection, in which all recyclates are collected mixed in with the rest of the waste, and the desired material is then sorted out and cleaned at a central sorting facility. This results in a large amount of recyclable waste, paper especially, being too soiled to reprocess, but has advantages as well: the city need not pay for a separate collection of recyclates and no public education is needed. Any changes to which materials are recyclable is easy to accommodate as all sorting happens in a central location.

In a commingled or single-stream system, all recyclables for collection are mixed but kept separate from other waste. This greatly reduces the need for post-collection cleaning but does require public education on what materials are recyclable.

Source separation is the other extreme, where each material is cleaned and sorted prior to collection. This method requires the least post-collection sorting and produces the purest recyclates, but incurs additional operating costs for collection of each separate material. An extensive public education program is also required, which must be successful if recyclate contamination is to be avoided.

Source separation used to be the preferred method due to the high sorting costs incurred by commingled (mixed waste) collection. Advances in sorting technology, however, have lowered this overhead substantially—many areas which had developed source separation programs have since switched to co-mingled collection.

Buy-back Centers

Buy-back centers differ in that the cleaned recyclates are purchased, thus providing a clear incentive for use and creating a stable supply. The post-processed material can then be sold. If this is profitable, this conserves the emission of greenhouse gases; if unprofitable, it increases the emission of greenhouse gasses. Government subsidies are necessary to make buy-back centres a viable enterprise. In 1993, according to the U.S. National Waste & Recycling Association, it costs on average US$50 to process a ton of material, which can be resold for US$30.

In the US, the value per ton of mixed recyclables was US$180 in 2011, US$80 in 2015, and US100 in 2017.

In 2017, glass is essentially valueless, because of the low cost of sand, its major component; low oil costs thwarts plastic recycling.

In 2017, Napa, California was reimbursed about 20% of its costs in recycling.

Drop-off Centers

Drop-off centers require the waste producer to carry the recyclates to a central location, either an installed or mobile collection station or the reprocessing plant itself. They are the easiest type of collection to establish but suffer from low and unpredictable throughput.

Distributed Recycling

For some waste materials such as plastic, recent technical devices called recyclebots enable a form of distributed recycling. Preliminary life-cycle analysis (LCA) indicates that such distributed recycling of HDPE to make filament of 3-D printers in rural regions is energetically favorable to either using virgin resin or conventional recycling processes because of reductions in transportation energy.

Sorting

Once commingled recyclates are collected and delivered to a central collection facility, the different types of materials must be sorted. This is done in a series of stages, many of which involve automated processes such that a truckload of material can be fully sorted in less than an hour. Some plants can now sort the materials automatically, known as single-stream recycling. In plants, a variety of materials is sorted such as paper, different types of plastics, glass, metals, food scraps, and most types of batteries. A 30 percent increase in recycling rates has been seen in the areas where these plants exist.

Initially, the commingled recyclates are removed from the collection vehicle and placed on a conveyor belt spread out in a single layer. Large pieces of corrugated fiberboard and plastic bags are removed by hand at this stage, as they can cause later machinery to jam.

Early sorting of recyclable materials: glass and plastic bottles in Poland

Next, automated machinery such as disk screens and air classifiers separate the recyclates by weight, splitting lighter paper and plastic from heavier glass and metal. Cardboard is removed from the mixed paper and the most common types of plastic, PET (#1) and HDPE (#2), are collected. This separation is usually done by hand but has become automated in some sorting centers: a spectroscopic scanner is used to differentiate between different types of paper and plastic based on the absorbed wavelengths, and subsequently divert each material into the proper collection channel.

Strong magnets are used to separate out ferrous metals, such as iron, steel, and tin cans. Non-ferrous metals are ejected by magnetic eddy currents in which a rotating magnetic field induces an electric current around the aluminum cans, which in turn creates a magnetic eddy current inside the cans. This magnetic eddy current is repulsed by a large magnetic field, and the cans are ejected from the rest of the recyclate stream.

A recycling point in New Byth, Scotland, with separate containers for paper, plastics, and differently colored glass

Finally, glass is sorted according to its color: brown, amber, green, or clear. It may either be sorted by hand, or via an automated machine that uses colored filters to detect different colors. Glass fragments smaller than 10 millimetres (0.39 in) across cannot be sorted automatically, and are mixed together as "glass fines."

This process of recycling as well as reusing the recycled material has proven advantageous because it reduces amount of waste sent to landfills, conserves natural resources, saves energy, reduces greenhouse gas emissions, and helps create new jobs. Recycled materials can also be converted into new products that can be consumed again, such as paper, plastic, and glass.

The City and County of San Francisco's Department of the Environment is attempting to achieve a citywide goal of generating zero waste by 2020. San Francisco's refuse hauler, Recology, operates an effective recyclables sorting facility in San Francisco, which helped San Francisco reach a record-breaking diversion rate of 80%.

Rinsing

Food packaging should no longer contain any organic matter (organic matter, if any, needs to be placed in a biodegradable waste bin or be buried in a garden). Since no trace of biodegradable material is best kept in the packaging before placing it in a trash bag, some packaging also needs to be rinsed.

Recycling Industrial Waste

Mounds of shredded rubber tires are ready for processing

Although many government programs are concentrated on recycling at home, a 64% of waste in the United Kingdom is generated by industry. The focus of many recycling programs done by industry is the cost–effectiveness of recycling. The ubiquitous nature of cardboard packaging makes cardboard a commonly recycled waste product by companies that deal heavily in packaged goods, like retail stores, warehouses, and distributors of goods. Other industries deal in niche or specialized products, depending on the nature of the waste materials that are present.

The glass, lumber, wood pulp and paper manufacturers all deal directly in commonly recycled materials; however, old rubber tires may be collected and recycled by independent tire dealers for a profit.

Levels of metals recycling are generally low. In 2010, the International Resource Panel, hosted by the United Nations Environment Programme (UNEP) published reports on metal stocks that exist within society and their recycling rates. The Panel reported that the increase in the use of metals during the 20th and into the 21st century has led to a substantial shift in metal stocks from below ground to use in applications within society above ground. For example, the in-use stock of copper in the USA grew from 73 to 238 kg per capita between 1932 and 1999.

The report authors observed that, as metals are inherently recyclable, the metal stocks in society can serve as huge mines above ground (the term "urban mining" has been coined with this idea in mind). However, they found that the recycling rates of many metals are very low. The report warned that the recycling rates of some rare metals used in applications such as mobile phones, battery packs for hybrid cars and fuel cells, are so low that unless future end-of-life recycling rates are dramatically stepped up these critical metals will become unavailable for use in modern technology.

The military recycles some metals. The U.S. Navy's Ship Disposal Program uses ship breaking to reclaim the steel of old vessels. Ships may also be sunk to create an artificial reef. Uranium is a very dense metal that has qualities superior to lead and titanium for many military and industrial uses. The uranium left over from processing it into nuclear weapons and fuel for nuclear reactors is called depleted uranium, and it is used by all branches of the U.S. military use for armour-piercing shells and shielding.

The construction industry may recycle concrete and old road surface pavement, selling their waste materials for profit.

Some industries, like the renewable energy industry and solar photovoltaic technology, in particular, are being proactive in setting up recycling policies even before there is considerable volume to their waste streams, anticipating future demand during their rapid growth.

Recycling of plastics is more difficult, as most programs are not able to reach the necessary level of quality. Recycling of PVC often results in downcycling of the material, which means only products of lower quality standard can be made with the recycled material. A new approach which allows an equal level of quality is the Vinyloop process. It was used after the London Olympics 2012 to fulfill the PVC Policy.

E-waste Recycling

E-waste is a growing problem, accounting for 20-50 million metric tons of global waste per year according to the EPA. It is also the fastest growing waste stream in the EU. Many recyclers do not recycle e-waste responsibly. After the cargo barge Khian Sea dumped 14,000 metric tons of toxic ash in Haiti, the Basel Convention was formed to stem the flow of hazardous substances into poorer countries. They created the e-Stewards certification to ensure that recyclers are held to the highest standards for environmental responsibility and to help consumers identify responsible recyclers. This works alongside other prominent legislation, such as the Waste Electrical and Electronic

Equipment Directive of the EU the United States National Computer Recycling Act, to prevent poisonous chemicals from entering waterways and the atmosphere.

Microprocessors retrieved from waste stream

In the recycling process, television sets, monitors, cell phones, and computers are typically tested for reuse and repaired. If broken, they may be disassembled for parts still having high value if labor is cheap enough. Other e-waste is shredded to pieces roughly 10 centimetres (3.9 in) in size, and manually checked to separate out toxic batteries and capacitors which contain poisonous metals. The remaining pieces are further shredded to 10 millimetres (0.39 in) particles and passed under a magnet to remove ferrous metals. An eddy current ejects non-ferrous metals, which are sorted by density either by a centrifuge or vibrating plates. Precious metals can be dissolved in acid, sorted, and smelted into ingots. The remaining glass and plastic fractions are separated by density and sold to re-processors. Television sets and monitors must be manually disassembled to remove lead from CRTs or the mercury backlight from LCDs.

Plastic Recycling

A container for recycling used plastic spoons into material for 3D printing

Plastic recycling is the process of recovering scrap or waste plastic and reprocessing the material into useful products, sometimes completely different in form from their original state. For instance, this could mean melting down soft drink bottles and then casting them as plastic chairs and tables.

Physical Recycling

Some plastics are remelted to form new plastic objects; for example, PET water bottles can be converted into polyester destined for clothing. A disadvantage of this type of recycling is that the molecular weight of the polymer can change further and the levels of unwanted substances in the plastic can increase with each remelt.

Chemical Recycling

For some polymers, it is possible to convert them back into monomers, for example, PET can be treated with an alcohol and a catalyst to form a dialkyl terephthalate. The terephthalate diester can be used with ethylene glycol to form a new polyester polymer, thus making it possible to use the pure polymer again.

Waste Plastic Pyrolysis to Fuel Oil

Another process involves conversion of assorted polymers into petroleum by a much less precise thermal depolymerization process. Such a process would be able to accept almost any polymer or mix of polymers, including thermoset materials such as vulcanized rubber tires and the biopolymers in feathers and other agricultural waste. Like natural petroleum, the chemicals produced can be used as fuels or as feedstock. A RESEM Technology plant of this type in Carthage, Missouri, USA, uses turkey waste as input material. Gasification is a similar process but is not technically recycling since polymers are not likely to become the result. Plastic Pyrolysis can convert petroleum based waste streams such as plastics into quality fuels, carbons. Given below is the list of suitable plastic raw materials for pyrolysis:

- Mixed plastic (HDPE, LDPE, PE, PP, Nylon, Teflon, PS, ABS, FRP, etc.)
- Mixed waste plastic from waste paper mill
- Multi-layered plastic

Recycling Codes

In order to meet recyclers' needs while providing manufacturers a consistent, uniform system, a coding system was developed. The recycling code for plastics was introduced in 1988 by the plastics industry through the Society of the Plastics Industry. Because municipal recycling programs traditionally have targeted packaging—primarily bottles and containers—the resin coding system offered a means of identifying the resin content of bottles and containers commonly found in the residential waste stream.

Plastic products are printed with numbers 1–7 depending on the type of resin. Type 1 (polyethylene terephthalate) is commonly found in soft drink and water bottles. Type 2 (high-density polyethylene) is found in most hard plastics such as milk jugs, laundry detergent bottles, and some dishware. Type 3 (polyvinyl chloride) includes items such as shampoo bottles, shower curtains, hula hoops, credit cards, wire jacketing, medical equipment, siding, and piping. Type 4 (low-density polyethylene) is found in shopping bags, squeezable bottles, tote bags, clothing, furniture, and carpet. Type 5 is polypropylene and makes up syrup bottles, straws, Tupperware, and some automotive parts. Type 6 is polystyrene and makes up meat trays, egg cartons, clamshell containers, and compact disc cases. Type 7 includes all other plastics such as bulletproof materials, 3- and 5-gallon water bottles, and sunglasses. Having a recycling code or the chasing arrows logo on a material is not an automatic indicator that a material is recyclable but rather an explanation of what the material is. Types 1 and 2 are the most commonly recycled.

Recycling codes on products

Economic Impact

Critics dispute the net economic and environmental benefits of recycling over its costs, and suggest that proponents of recycling often make matters worse and suffer from confirmation bias. Specifically, critics argue that the costs and energy used in collection and transportation detract from (and outweigh) the costs and energy saved in the production process; also that the jobs produced by the recycling industry can be a poor trade for the jobs lost in logging, mining, and other industries associated with production; and that materials such as paper pulp can only be recycled a few times before material degradation prevents further recycling.

The National Waste and Recycling Association (NWRA), reported in May 2015, that recycling and waste made a $6.7 billion economic impact in Ohio, U.S., and employed 14,000 people.

Cost–benefit Analysis

There is some debate over whether recycling is economically efficient. It is said that dumping 10,000 tons of waste in a landfill creates six jobs while recycling 10,000 tons of waste can create over 36 jobs. However, the cost effectiveness of creating the additional jobs remains unproven. According to the U.S. Recycling Economic Informational Study, there are over 50,000 recycling establishments that have created over a million jobs in the US. Two years after New York City declared that implementing recycling programs would be "a drain on the city," New York City leaders realized that an efficient recycling system could save the city over $20 million. Municipalities often see fiscal benefits from implementing recycling programs, largely due to the reduced landfill costs. A study conducted by the Technical University of Denmark according to the Economist found that in 83 percent of cases, recycling is the most efficient method to dispose of household waste. However, a 2004 assessment by the Danish Environmental Assessment Institute concluded that incineration was the most effective method for disposing of drink containers, even aluminium ones.

Environmental effects of recycling		
Material	Energy savings	Air pollution savings
Aluminium	95%	95%
Cardboard	24%	—
Glass	5–30%	20%
Paper	40%	73%
Plastics	70%	—
Steel	60%	—

Fiscal efficiency is separate from economic efficiency. Economic analysis of recycling does not include what economists call externalities, which are unpriced costs and benefits that accrue to individuals outside of private transactions. Examples include: decreased air pollution and greenhouse gases from incineration, reduced hazardous waste leaching from landfills, reduced energy consumption, and reduced waste and resource consumption, which leads to a reduction in environmentally damaging mining and timber activity. About 4,000 minerals are known, of these only a few hundred minerals in the world are relatively common. Known reserves of phosphorus will be exhausted within the next 100 years at current rates of usage. Without mechanisms such as taxes or subsidies to internalize externalities, businesses may ignore them despite the costs imposed on society. To make such nonfiscal benefits economically relevant, advocates have pushed for legislative action to increase the demand for recycled materials. The United States Environmental Protection Agency (EPA) has concluded in favor of recycling, saying that recycling efforts reduced the country's carbon emissions by a net 49 million metric tonnes in 2005. In the United Kingdom, the Waste and Resources Action Programme stated that Great Britain's recycling efforts reduce CO_2 emissions by 10–15 million tonnes a year. Recycling is more efficient in densely populated areas, as there are economies of scale involved.

Certain requirements must be met for recycling to be economically feasible and environmentally effective. These include an adequate source of recyclates, a system to extract those recyclates from the waste stream, a nearby factory capable of reprocessing the recyclates, and a potential demand for the recycled products. These last two requirements are often overlooked—without both an industrial market for production using the collected materials and a consumer market for the manufactured goods, recycling is incomplete and in fact only "collection".

Wrecked automobiles gathered for smelting

Free-market economist Julian Simon remarked "There are three ways society can organize waste disposal: (a) commanding, (b) guiding by tax and subsidy, and (c) leaving it to the individual and the market". These principles appear to divide economic thinkers today.

Frank Ackerman favours a high level of government intervention to provide recycling services. He believes that recycling's benefit cannot be effectively quantified by traditional *laissez-faire* economics. Allen Hershkowitz supports intervention, saying that it is a public service equal to education and policing. He argues that manufacturers should shoulder more of the burden of waste disposal.

Paul Calcott and Margaret Walls advocate the second option. A deposit refund scheme and a small refuse charge would encourage recycling but not at the expense of fly-tipping. Thomas C. Kinnaman concludes that a landfill tax would force consumers, companies and councils to recycle more.

Most free-market thinkers detest subsidy and intervention because they waste resources. Terry Anderson and Donald Leal think that all recycling programmes should be privately operated, and therefore would only operate if the money saved by recycling exceeds its costs. Daniel K. Benjamin argues that it wastes people's resources and lowers the wealth of a population.

Trade in Recyclates

Certain countries trade in unprocessed recyclates. Some have complained that the ultimate fate of recyclates sold to another country is unknown and they may end up in landfills instead of reprocessed. According to one report, in America, 50–80 percent of computers destined for recycling are actually not recycled. There are reports of illegal-waste imports to China being dismantled and recycled solely for monetary gain, without consideration for workers' health or environmental damage. Although the Chinese government has banned these practices, it has not been able to eradicate them. In 2008, the prices of recyclable waste plummeted before rebounding in 2009. Cardboard averaged about £53/tonne from 2004–2008, dropped to £19/tonne, and then went up to £59/tonne in May 2009. PET plastic averaged about £156/tonne, dropped to £75/tonne and then moved up to £195/tonne in May 2009.

Certain regions have difficulty using or exporting as much of a material as they recycle. This problem is most prevalent with glass: both Britain and the U.S. import large quantities of wine bottled in green glass. Though much of this glass is sent to be recycled, outside the American Midwest there is not enough wine production to use all of the reprocessed material. The extra must be downcycled into building materials or re-inserted into the regular waste stream.

Similarly, the northwestern United States has difficulty finding markets for recycled newspaper, given the large number of pulp mills in the region as well as the proximity to Asian markets. In other areas of the U.S., however, demand for used newsprint has seen wide fluctuation.

In some U.S. states, a program called RecycleBank pays people to recycle, receiving money from local municipalities for the reduction in landfill space which must be purchased. It uses a single stream process in which all material is automatically sorted.

Criticisms and Responses

Much of the difficulty inherent in recycling comes from the fact that most products are not designed with recycling in mind. The concept of sustainable design aims to solve this problem, and was laid out in the book *Cradle to Cradle: Remaking the Way We Make Things* by architect William McDonough and chemist Michael Braungart. They suggest that every product (and all packaging they require) should have a complete "closed-loop" cycle mapped out for each component—a way in which every component will either return to the natural ecosystem through biodegradation or be recycled indefinitely.

Complete recycling is impossible from a practical standpoint. In summary, substitution and recycling strategies only delay the depletion of non-renewable stocks and therefore may buy time in the transition to true or strong sustainability, which ultimately is only guaranteed in an economy based on renewable resources.

—M. H. Huesemann, 2003

While recycling diverts waste from entering directly into landfill sites, current recycling misses the dissipative components. Complete recycling is impracticable as highly dispersed wastes become so diluted that the energy needed for their recovery becomes increasingly excessive. "For example, how will it ever be possible to recycle the numerous chlorinated organic hydrocarbons that have bioaccumulated in animal and human tissues across the globe, the copper dispersed in fungicides, the lead in widely applied paints, or the zinc oxides present in the finely dispersed rubber powder that is abraded from automobile tires?"

As with environmental economics, care must be taken to ensure a complete view of the costs and benefits involved. For example, paperboard packaging for food products is more easily recycled than most plastic, but is heavier to ship and may result in more waste from spoilage.

Energy and Material Flows

Bales of crushed steel ready for transport to the smelter

The amount of energy saved through recycling depends upon the material being recycled and the type of energy accounting that is used. Correct accounting for this saved energy can be accomplished with life-cycle analysis using real energy values. In addition, exergy, which is a measure of useful energy can be used. In general, it takes far less energy to produce a unit mass of recycled materials than it does to make the same mass of virgin materials.

Some scholars use emergy (spelled with an m) analysis, for example, budgets for the amount of energy of one kind (exergy) that is required to make or transform things into another kind of product or service. Emergy calculations take into account economics which can alter pure physics based results. Using emergy life-cycle analysis researchers have concluded that materials with large refining costs have the greatest potential for high recycle benefits. Moreover, the highest emergy efficiency accrues from systems geared toward material recycling, where materials are engineered to recycle back into their original form and purpose, followed by adaptive reuse systems where the materials are recycled into a different kind of product, and then by-product reuse systems where parts of the products are used to make an entirely different product.

The Energy Information Administration (EIA) states on its website that "a paper mill uses 40 percent less energy to make paper from recycled paper than it does to make paper from fresh lumber." Some critics argue that it takes more energy to produce recycled products than it does to dispose of them in traditional landfill methods, since the curbside collection of recyclables often requires a second waste truck. However, recycling proponents point out that a second timber or logging truck is eliminated when paper is collected for recycling, so the net energy consumption is the same. An Emergy life-cycle analysis on recycling revealed that fly ash, aluminum, recycled concrete aggregate, recycled plastic, and steel yield higher efficiency ratios, whereas the recycling of lumber generates the lowest recycle benefit ratio. Hence, the specific nature of the recycling process, the methods used to analyse the process, and the products involved affect the energy savings budgets.

It is difficult to determine the amount of energy consumed or produced in waste disposal processes in broader ecological terms, where causal relations dissipate into complex networks of material and energy flow. For example, "cities do not follow all the strategies of ecosystem development. Biogeochemical paths become fairly straight relative to wild ecosystems, with very reduced recycling, resulting in large flows of waste and low total energy efficiencies. By contrast, in wild ecosystems, one population's wastes are another population's resources, and succession results in efficient exploitation of available resources. However, even modernized cities may still be in the earliest stages of a succession that may take centuries or millennia to complete." How much energy is used in recycling also depends on the type of material being recycled and the process used to do so. Aluminium is generally agreed to use far less energy when recycled rather than being produced from scratch. The EPA states that "recycling aluminum cans, for example, saves 95 percent of the energy required to make the same amount of aluminum from its virgin source, bauxite." In 2009, more than half of all aluminium cans produced came from recycled aluminium.

Every year, millions of tons of materials are being exploited from the earth's crust, and processed into consumer and capital goods. After decades to centuries, most of these materials are "lost". With the exception of some pieces of art or religious relics, they are no longer engaged in the consumption process. Where are they? Recycling is only an intermediate solution for such materials, although it does prolong the residence time in the anthroposphere. For thermodynamic reasons, however, recycling cannot prevent the final need for an ultimate sink.

—*P. H. Brunner*

Economist Steven Landsburg has suggested that the sole benefit of reducing landfill space is trumped by the energy needed and resulting pollution from the recycling process. Others, however, have calculated through life-cycle assessment that producing recycled paper uses less energy and water than harvesting, pulping, processing, and transporting virgin trees. When less recycled paper is used, additional energy is needed to create and maintain farmed forests until these forests are as self-sustainable as virgin forests.

Other studies have shown that recycling in itself is inefficient to perform the "decoupling" of economic development from the depletion of non-renewable raw materials that is necessary for sustainable development. The international transportation or recycle material flows through " different trade networks of the three countries result in different flows, decay rates, and potential recycling returns."[1] As global consumption of a natural resources grows, its depletion is inevitable. The best recycling can do is to delay, complete closure of material loops to achieve 100 percent recycling of nonrenewables is impossible as micro-trace materials dissipate into the environment causing severe damage to the planet's ecosystems. Historically, this was identified as the metabolic rift by Karl Marx, who identified the unequal exchange rate between energy and nutrients flowing from rural areas to feed urban cities that create effluent wastes degrading the planet's ecological capital, such as loss in soil nutrient production. Energy conservation also leads to what is known as Jevon's paradox, where improvements in energy efficiency lowers the cost of production and leads to a rebound effect where rates of consumption and economic growth increases.

A shop in New York only sells items recycled from demolished buildings

Costs

The amount of money actually saved through recycling depends on the efficiency of the recycling program used to do it. The Institute for Local Self-Reliance argues that the cost of recycling depends on various factors, such as landfill fees and the amount of disposal that the community recycles. It states that communities begin to save money when they treat recycling as a replacement for their traditional waste system rather than an add-on to it and by "redesigning their collection schedules and/or trucks."

In some cases, the cost of recyclable materials also exceeds the cost of raw materials. Virgin plastic resin costs 40 percent less than recycled resin. Additionally, a United States Environmental Protection Agency (EPA) study that tracked the price of clear glass from July 15 to August 2, 1991, found that the average cost per ton ranged from $40 to $60 while a USGS report shows that the cost per ton of raw silica sand from years 1993 to 1997 fell between $17.33 and $18.10.

Comparing the market cost of recyclable material with the cost of new raw materials ignores economic externalities—the costs that are currently not counted by the market. Creating a new piece of plastic, for instance, may cause more pollution and be less sustainable than recycling a similar piece of plastic, but these factors will not be counted in market cost. A life cycle assessment can be used to determine the levels of externalities and decide whether the recycling may be worthwhile despite unfavorable market costs. Alternatively, legal means (such as a carbon tax) can be used to bring externalities into the market, so that the market cost of the material becomes close to the true cost.

Working Conditions

People in Brazil who earn their living by collecting and sorting garbage and selling them for recycling

The recycling of waste electrical and electronic equipment in India and China generates a significant amount of pollution. Informal recycling in an underground economy of these countries has generated an environmental and health disaster. High levels of lead (Pb), polybrominated diphenylethers (PBDEs), polychlorinated dioxins and furans, as well as polybrominated dioxins and furans (PCDD/Fs and PBDD/Fs) concentrated in the air, bottom ash, dust, soil, water, and sediments in areas surrounding recycling sites.

Environmental Impact

Economist Steven Landsburg, author of a paper entitled "Why I Am Not an Environmentalist," claimed that paper recycling actually reduces tree populations. He argues that because paper companies have incentives to replenish their forests, large demands for paper lead to large forests while reduced demand for paper leads to fewer "farmed" forests.

When foresting companies cut down trees, more are planted in their place. Most paper comes from pulp forests grown specifically for paper production. Many environmentalists point out, however, that "farmed" forests are inferior to virgin forests in several ways. Farmed forests are not able to fix the soil as quickly as virgin forests, causing widespread soil erosion and often requiring large amounts of fertilizer to maintain while containing little tree and wild-life biodiversity compared to virgin forests. Also, the new trees planted are not as big as the trees that were cut down, and the argument that there will be "more trees" is not compelling to forestry advocates when they are counting saplings.

In particular, wood from tropical rainforests is rarely harvested for paper because of their heterogeneity. According to the United Nations Framework Convention on Climate Change secretariat, the overwhelming direct cause of deforestation is subsistence farming (48% of deforestation) and commercial agriculture (32%), which is linked to food, not paper production.

Possible Income Loss and Social Costs

In some countries, recycling is performed by the entrepreneurial poor such as the karung guni, zabbaleen, the rag-and-bone man, waste picker, and junk man. With the creation of large recycling organizations that may be profitable, either by law or economies of scale, the poor are more likely to be driven out of the recycling and the remanufacturing market. To compensate for this loss of income, a society may need to create additional forms of societal programs to help support the poor. Like the parable of the broken window, there is a net loss to the poor and possibly the whole of a society to make recycling artificially profitable e.g. through the law. However, in Brazil and Argentina, waste pickers/informal recyclers work alongside the authorities, in fully or semi-funded cooperatives, allowing informal recycling to be legitimized as a paid public sector job.

Because the social support of a country is likely to be less than the loss of income to the poor undertaking recycling, there is a greater chance the poor will come in conflict with the large recycling organizations. This means fewer people can decide if certain waste is more economically reusable in its current form rather than being reprocessed. Contrasted to the recycling poor, the efficiency of their recycling may actually be higher for some materials because individuals have greater control over what is considered "waste."

One labor-intensive underused waste is electronic and computer waste. Because this waste may still be functional and wanted mostly by those on lower incomes, who may sell or use it at a greater efficiency than large recyclers.

Some recycling advocates believe that laissez-faire individual-based recycling does not cover all of society's recycling needs. Thus, it does not negate the need for an organized recycling program. Local government can consider the activities of the recycling poor as contributing to property blight.

Public Participation Rates

Changes that have been demonstrated to increase recycling rates include:

- Single-stream recycling

- Pay as you throw fees for trash

"Between 1960 and 2000, the world production of plastic resins increased 25-fold, while recovery of the material remained below 5 percent." Many studies have addressed recycling behaviour and strategies to encourage community involvement in recycling programmes. It has been argued that recycling behaviour is not natural because it requires a focus and appreciation for long-term planning, whereas humans have evolved to be sensitive to short-term survival goals; and that to overcome this innate predisposition, the best solution would be to use social pressure to compel participation in recycling programmes. However, recent studies have concluded that social pressure is unviable in this context. One reason for this is that social pressure functions well in small group sizes of 50 to 150 individuals (common to nomadic hunter–gatherer peoples) but not in communities numbering in the millions, as we see today. Another reason is that individual recycling does not take place in the public view.

In a study done by social psychologist Shawn Burn, it was found that personal contact with individuals within a neighborhood is the most effective way to increase recycling within a community. In his study, he had 10 block leaders talk to their neighbors and persuade them to recycle. A comparison group was sent fliers promoting recycling. It was found that the neighbors that were personally contacted by their block leaders recycled much more than the group without personal contact. As a result of this study, Shawn Burn believes that personal contact within a small group of people is an important factor in encouraging recycling. Another study done by Stuart Oskamp examines the effect of neighbors and friends on recycling. It was found in his studies that people who had friends and neighbors that recycled were much more likely to also recycle than those who didn't have friends and neighbors that recycled.

Many schools have created recycling awareness clubs in order to give young students an insight on recycling. These schools believe that the clubs actually encourage students to not only recycle at school but at home as well.

References

- "New Global Climate Prosperity Scoreboard Finds Over $1 Trillion Invested in Green Since 2007". Green Money Journal. 2010. Retrieved 11 June 2010

- Baccini and Bader 1996, 'Regionaler Stoffhaushalt' (Regional metabolism), Spektrum Akademischer Verlag, Heidelberg (Germany), ISBN 3-86025-235-6

- "Impact of closed-loop network configurations on carbon footprints: A case study in copiers". Citeulike.org. Retrieved 2014-06-08

- Marina Fischer-Kowalski, The Intellectual History of Materials Flow Analysis, Part I, 1860-1970, Journal of Industrial Ecology 2(1), 1998, pp 61-78, doi:10.1162/jiec.1998.2.1.61

- Ivanov D.A. "Supply chain multi-structural (re)-design." Archived August 12, 2011, at the Wayback Machine., International Journal of Integrated Supply Management, No. 5(1), 19-37., 2009

- Nakamura, S.; Kondo, Y. (2009). Waste Input-Output Analysis. Concepts and Application to Industrial Ecology. Springer. ISBN 978-1-4020-9901-4

- Daniel B. Müller, Stock dynamics for forecasting material flows--Case study for housing in The Netherlands, Ecological Economics 59(1), 2006, pp 142-156, doi:10.1016/j.ecolecon.2005.09.025

- Mentzer, J.T.; et al. (2001). "Defining Supply Chain Management". Journal of Business Logistics. 22 (2): 1–25. doi:10.1002/j.2158-1592.2001.tb00001

- Robert B. Handfield; Ernest L. Nichols (1999). Introduction to Supply Chain Management. New York: Prentice-Hall. p. 2. ISBN 0-13-621616-1

- "Regulatory Policy Center — Property Matters — James V. DeLong". Archived from the original on 14 April 2008. Retrieved February 28, 2008

- Todo, Y.; Matous, P.; Inoue, H. (11 July 2016). "The strength of long ties and the weakness of strong ties: Knowledge diffusion through supply chain networks". Research Policy. doi:10.1016/j.respol.2016.06.008

- Johnson, M. R. & McCarthy I. P. (2014) Product Recovery Decisions within the Context of Extended Producer Responsibility. Journal of Engineering and Technology Management 34, 9-28

- Betty A. Kildow (2011), Supply Chain Management Guide to Business Continuity, American Management Association, ISBN 9780814416457

- Hogye, Thomas Q. "The Anatomy of a Computer Recycling Process" (PDF). California Department of Resources Recycling and Recovery. Retrieved October 13, 2014

- "The Relational View: Cooperative Strategy and Sources of Interorganizational Competitive Advantage". The Academy of Management Review. 23: 660. doi:10.2307/259056

- Baechler, Christian; DeVuono, Matthew; Pearce, Joshua M. (2013). "Distributed Recycling of Waste Polymer into RepRap Feedstock". Rapid Prototyping Journal. 19 (2): 118–125. doi:10.1108/13552541311302978

- Cleveland, Cutler J.; Morris, Christopher G. (November 15, 2013). Handbook of Energy: Chronologies, Top Ten Lists, and Word Clouds. Elsevier. p. 461. ISBN 978-0-12-417019-3

- Regulatory Policy Center WASTING AWAY: Mismanaging Municipal Solid Waste Archived 14 April 2008 at the Wayback Machine.. Retrieved November 4, 2006

- Zaman, A. U.; Lehmann, S. (2011). "Challenges and opportunities in transforming a city into a "Zero Waste City"". Challenges. 2 (4): 73–93. doi:10.3390/challe2040073

- Burn, Shawn. "Social Psychology and the Stimulation of Recycling Behaviors: The Block Leader Approach." Journal of Applied Social Psychology 21.8 (2006): 611–629

- Lynn R. Kahle; Eda Gurel-Atay, eds. (2014). Communicating Sustainability for the Green Economy. New York: M.E. Sharpe. ISBN 978-0-7656-3680-5

Permissions

All chapters in this book are published with permission under the Creative Commons Attribution Share Alike License or equivalent. Every chapter published in this book has been scrutinized by our experts. Their significance has been extensively debated. The topics covered herein carry significant information for a comprehensive understanding. They may even be implemented as practical applications or may be referred to as a beginning point for further studies.

We would like to thank the editorial team for lending their expertise to make the book truly unique. They have played a crucial role in the development of this book. Without their invaluable contributions this book wouldn't have been possible. They have made vital efforts to compile up to date information on the varied aspects of this subject to make this book a valuable addition to the collection of many professionals and students.

This book was conceptualized with the vision of imparting up-to-date and integrated information in this field. To ensure the same, a matchless editorial board was set up. Every individual on the board went through rigorous rounds of assessment to prove their worth. After which they invested a large part of their time researching and compiling the most relevant data for our readers.

The editorial board has been involved in producing this book since its inception. They have spent rigorous hours researching and exploring the diverse topics which have resulted in the successful publishing of this book. They have passed on their knowledge of decades through this book. To expedite this challenging task, the publisher supported the team at every step. A small team of assistant editors was also appointed to further simplify the editing procedure and attain best results for the readers.

Apart from the editorial board, the designing team has also invested a significant amount of their time in understanding the subject and creating the most relevant covers. They scrutinized every image to scout for the most suitable representation of the subject and create an appropriate cover for the book.

The publishing team has been an ardent support to the editorial, designing and production team. Their endless efforts to recruit the best for this project, has resulted in the accomplishment of this book. They are a veteran in the field of academics and their pool of knowledge is as vast as their experience in printing. Their expertise and guidance has proved useful at every step. Their uncompromising quality standards have made this book an exceptional effort. Their encouragement from time to time has been an inspiration for everyone.

The publisher and the editorial board hope that this book will prove to be a valuable piece of knowledge for students, practitioners and scholars across the globe.

Index

A
Active System, 59

B
Biosphere Rules, 53-54, 126, 129
Blue Economy, 123, 129
Business Process Integration, 166
Buy-back Centers, 200-201

C
Circular Business Model, 131, 138
Circular Economy, 3, 13, 56, 120, 123-131, 137-140, 181-182
Closed Loop Production, 23, 53, 124, 129
Community Organization, 77, 79
Corporate Planning, 76
Cradle-to-gate, 23-24, 43
Cradle-to-grave, 21, 23, 82

D
De-linking, 97
Deconstruction, 89-90, 179
Dematerialization (economics), 120
Design for the Environment, 2, 16, 28-30, 32
Digital Technology, 127
Distributed Recycling, 201, 217
Drop-off Centers, 200-201

E
Eco-costs Value Ratio, 73, 93, 133
Eco-efficiency, 1-2, 29, 39, 61, 72, 98, 119-120, 129, 132-135, 137-138
Eco-efficient Value Creation, 93, 98-99
Eco-industrial Park, 3-4, 6, 11, 60-61
Eco-investing, 136-137
Ecodesign, 43-46, 56-58
Ecological Footprint, 62, 73, 100-106, 115, 118-119
Ecological Modernization, 44, 66-69, 72
Ecologically Based Lca, 25
Ecomechatronics, 44-45
Economy Wide Effects, 113
Energy Accounting, 73, 107, 211
Energy Balance, 107
Energy Management, 47, 107
Energy Production, 4, 23, 26, 104, 106
Energy Recovery, 26, 182
Environmental Benchmarking, 99-100
Environmental Effect Analysis, 58
Environmental Full-cost Accounting, 114
Environmental Management System, 6, 39, 44, 63-64, 66
Exergy Based Lca, 25
Extended Producer Responsibility, 2, 16, 33-34, 36-37, 39, 43, 91, 217

F
Food Systems, 71
Full-cost Accounting, 114-116
Future Outlays, 115-116

G
Geospatial Models, 158
Globalization Era, 164

H
Helix of Sustainability, 16, 40-41
Hidden Costs, 115

I
Income Level Variation, 113
Industrial Metabolism, 1, 10-11, 66-67, 139-140, 176-177
Industrial Organization, 9, 120-121, 123
Industrial Symbiosis, 2, 7-9, 12-15, 60-61, 67
Integrated Chain Management, 44, 62
Integration Era, 164
Internalized Environmental Costs, 186

K
Khazzoom-brookes Postulate, 109-110

L
Life Cycle Energy Analysis, 25
Life Cycle Inventory, 19, 21, 23, 42-43, 143

Life-cycle Assessment, 16, 48, 133, 212
Linear Model, 82, 124

M
Machine Components, 46
Machine Control, 46-47
Material Flow, 6, 139-140, 143-145, 147-148, 150
Material Processing, 5
Model Animation, 157
Multimethod Simulation Modeling, 154

N
National Ambient Air Quality Standards (NAAQS), 32

O
Outsourced Manufacturing, 164-165

P
Package Deposit Programs, 182
Passive Systems, 59
Performance Measurement, 123, 169
Physical Distribution, 169
Plant Simulation, 139, 147-150
Plastic Bags, 36, 202
Printer Cartridges, 184
Process Balance, 142

R
Raw Material Extraction, 5, 17, 29
Rebound Effect, 73, 99, 108-114, 213
Recyclates, 196-202, 208-210
Recycling Industrial Waste, 203
Reduce, 4-5, 14, 27-28, 30-31, 34, 36-37, 58-61, 63-64, 66-67, 81-83, 87, 89, 91-92, 94, 98, 110, 112-114, 162, 167-168, 170, 172, 192-193, 198-199, 208
Refilling Programs, 184

Refuse, 87, 196, 203, 209
Regenerative Design, 44, 47, 69, 71
Regenerative Versus Sustainable, 70
Regifting, 184
Remanufacturing, 33, 123, 130, 182, 187-192, 215
Repurposing, 177, 180, 184-185, 187
Reuse, 4-5, 8, 13, 30, 34-36, 38, 41-42, 51, 54, 57-58, 61, 81, 83-92, 119, 123, 126, 129, 147, 177-187, 192-193, 195, 205, 211

S
Social Responsibility, 24, 115, 172
Specialization Era, 164-165
Strategy Building, 75
Stratospheric Ozone Protection, 33
Substance Flow Analysis, 140, 145
Supply Chain Management, 62-63, 66, 139, 160-166, 170-171, 175-176, 217
Synthetic Materials, 49-51

T
Take-back, 34-36, 91, 108
Toxic Substances, 33

V
Virtual Exchange, 179, 185

W
Waste Exchanges, 185
Waste Management, 4, 6, 15, 22, 34-35, 37, 39, 81-82, 84, 88, 91-92, 116-118, 131, 181
Waste Water, 140, 146, 185
Well-to-wheel, 24

Z
Zero Waste Hierarchy, 91-92
Zero Waste Jurisdictions, 92